Ben Stacy Jerrik (Ed.)

Brington, Northamptonshire

Ben Stacy Jerrik (Ed.)

Brington, Northamptonshire

Wanderers F.C. Daventry District, Henry Holmes Stewart

Part Press

Contents

Brington,_Northamptonshire

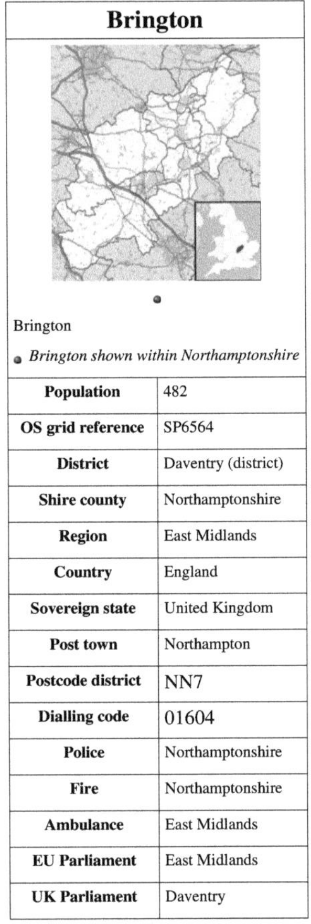

Brington	
Brington	
Brington shown within Northamptonshire	
Population	482
OS grid reference	SP6564
District	Daventry (district)
Shire county	Northamptonshire
Region	East Midlands
Country	England
Sovereign state	United Kingdom
Post town	Northampton
Postcode district	NN7
Dialling code	01604
Police	Northamptonshire
Fire	Northamptonshire
Ambulance	East Midlands
EU Parliament	East Midlands
UK Parliament	Daventry

Brington is a civil parish in the Daventry district of the county of Northamptonshire in England. At the time of the 2001 census, the parish population was 482 people.[1]

It contains three villages:

- Great Brington
- Little Brington
- Nobottle

Notable people

Rev. Henry Holmes Stewart (1847–1937), who won the FA Cup with Wanderers in 1873, was rector at the parish church from 1878 to 1898.[2]

References

[1] Office of National Statistics: Brington CP: Parish headcounts (http://neighbourhood.statistics.gov.uk/dissemination/LeadTableView.do?a=7&b=797165&c=brington&d=16&e=15&g=472326&i=1001x1003x1004&m=0&r=1&s=1257592443625&enc=1&dsFamilyId=779). Retrieved 7 November 2009

[2] Venn, J.; Venn, J. A., eds. (1922–1958). " Stewart, Henry (http://venn.lib.cam.ac.uk/cgi-bin/search.pl?sur=&suro=c&fir=&firo=c&cit=&cito=c&c=all&tex=STWT866HH&sye=&eye=&col=all&maxcount=50)". *Alumni Cantabrigienses (10 vols)* (online ed.). Cambridge University Press.

Wanderers_F.C.

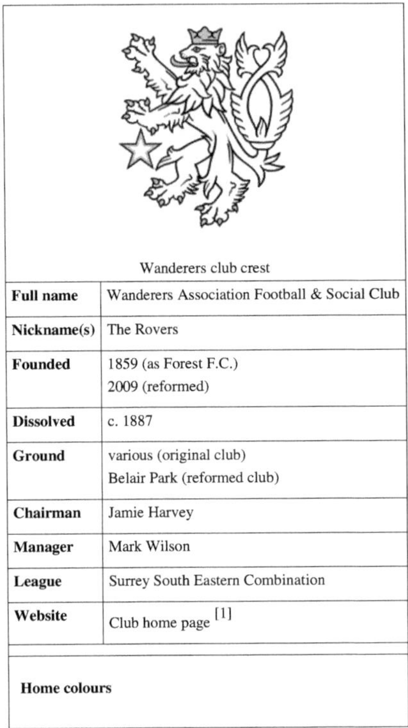

Wanderers club crest

Full name	Wanderers Association Football & Social Club
Nickname(s)	The Rovers
Founded	1859 (as Forest F.C.) 2009 (reformed)
Dissolved	c. 1887
Ground	various (original club) Belair Park (reformed club)
Chairman	Jamie Harvey
Manager	Mark Wilson
League	Surrey South Eastern Combination
Website	Club home page [1]

Home colours

Wanderers Football Club is an English amateur football club, based in London. Founded as **Forest Football Club** in 1859, the club changed its name to Wanderers in 1864. Comprising mainly former pupils of the leading English public schools, Wanderers was among the dominant teams of the early years of organised football and won the Football Association Challenge Cup (known in the modern era as the FA Cup) on five occasions, including defeating Royal Engineers in the first FA Cup final in 1872.

The club played only friendly matches until the advent of the FA Cup in 1871, with the rules often differing from match to match as various sets of rules were in use at the time. Even after the formation of The Football Association (the FA) in 1863, of which the club was among the founder members, Wanderers continued to play matches under other rules, but became one of the strongest teams playing by FA rules. They won the FA Cup three times in succession during the late 1870s, a feat which has only been repeated once. Among the players who represented the club were C. W. Alcock, the so-called "father of modern sport", and A.F. Kinnaird, regarded as the greatest player of his day. In keeping with its name, the club never had a home stadium of its own but played at various locations in London and the surrounding area. By the 1880s the club's fortunes had declined and it was reduced to playing only an annual match against Harrow School, the *alma mater* of many of its founders.

The club was reformed, with the endorsement of the descendants of the Alcock family, for the purposes of fundraising for UNICEF UK and the revived club now competes in the Surrey South Eastern Combination.

History

Early years (1859–1871)

The club was initially formed as Forest Football Club in 1859 by a number of former public school pupils, including C. W. Alcock, who had finished his education at Harrow School in the same year.[2] The other founders were Alcock's brother, John F. Alcock, J. Pardoe and brothers A. and W. J. Thompson.[3]

For the first two years of the club's existence, the players organised matches among themselves at Snaresbrook near Epping Forest, possibly on land owned by the Earl of Mornington.[4] Forest's first match against another club took place on 15 March 1862, and resulted in a victory over Crystal Palace (not the modern club of the same name).[5] Both this match, and a return fixture between the two teams the following month, involved fifteen players on each team.[6] At the time, the rules of association football

The only known photo of the team, taken in 1863 when the club was still known as Forest F.C.

had not been codified, and many variants existed, differing in the number of players per team, whether players were permitted to play the ball with their hands, or the method of scoring goals. In 1863 the Forest club was among the founder members of The Football Association (the FA) and adopted the rules set down by that body, although they continued to play occasional matches under other sets of rules against clubs not affiliated to the FA.[7]

The following year, the club played its first match under the name Wanderers Football Club, against No Names Club of Kilburn.[8] Alcock had decided, possibly because of the expense the club was incurring by owning its own ground, to turn it into a "wandering" team with no fixed home venue, however it appears that some of the club's members opposed this idea.[9] For the following season teams operated under both names, with several players appearing for both, and indeed Forest and Wanderers even played each other in one match, but after 1865 there is no record of any further matches under the Forest name.[10] The Wanderers initially fared well, losing only one of their sixteen matches in the 1865–66 season, but over the subsequent four seasons the team's fortunes declined significantly and Alcock also found it increasingly difficult to ensure that eleven of his players actually turned up for a match, with the club often forced to play with fewer than the required number of players or borrow some from their opponents.[11] During this period the club played a number of "home" matches at Battersea Park and Middlesex County Cricket Club's Lillie Bridge Grounds.[12] Wanderers subsequently made Kennington Oval its semi-permanent home in 1869.[13] The club played 151 matches at The Oval.

Cup success (1872–1878)

In the 1870–71 season, the Wanderers finally turned around their fortunes, losing only five of thirty-seven matches played.[14] For the following season the FA, following a suggestion by Alcock, initiated the Football Association Challenge Cup, a knock-out tournament open to all member clubs. Due to a combination of their opponents withdrawing and an unusual rule in place at the time which allowed both clubs to progress to the next round in the event of a draw, Wanderers only won one game in the four rounds leading up to the final, held at the Kennington Oval on 16 March 1872. The club beat the Royal Engineers 1–0 to become the first ever winners of the cup, the winning goal being scored by Morton Betts, who was playing under the pseudonym "A. H. Chequer".[15] The following season, under the competition's original rules, Wanderers, as holders, received a bye all the way to the final. In the final Wanderers beat Oxford University 2–0 to retain the cup, thanks in large part to the performance of A. F. Kinnaird.[16] The club was unable to replicate this success over the next two seasons, although the team did manage a club record 16–0 victory over Farningham in the first round of the 1874–75 FA Cup.[17]

The programme from Wanderers' match away to Queen's Park in October 1875.

In October 1875, Wanderers travelled to Scotland for the first time, to play a match against the leading team from north of the border, Queen's Park. Despite fielding their strongest team, Wanderers were outclassed by the Scots and lost 5–0. The London club gained its revenge four months later, however, when Queen's Park travelled to London for a re-match and lost 2–0. This was the first match the Glasgow club, which had been formed nine years earlier, had ever lost.[18] Wanderers reached the semi-finals of the FA Cup without conceding a goal and then defeated Swifts to set up a final against Old Etonians. The Etonians' team contained five former Wanderers players, including Kinnaird.[19] After the initial match finished in a 1–1 draw, Wanderers won the replay 3–0 to win the tournament for the third time.

The following season, with Kinnaird back in the team, Wanderers overcame indifferent early form to again reach the Cup final, and defeated Oxford University to retain the trophy. Wanderers again dominated the competition in the 1877–78 season, scoring nine goals in both their first and second round matches. The final was a rematch of the 1872 final and Wanderers again defeated Royal Engineers to win an unprecedented third consecutive FA Cup. The rules of the competition stated that under such circumstances the trophy would be retired and become the permanent property of the victorious club, but Alcock returned the cup to the FA on the condition that the rule be removed and no other team permitted to claim it on a permanent basis. Following the final, Wanderers played the reigning Scottish Cup holders, Vale of Leven, but lost 3–1.[20]

Decline (1879–1887)

The Wanderers' fortunes declined rapidly following the club's hat-trick of FA Cup wins. By 1878, football clubs had been set up for former pupils of all the leading public schools, and many leading players chose to play for their respective old boys' team instead. Wanderers' fixture list was dramatically reduced in the 1878–79 season, and the team was knocked out of the FA Cup in the first round, losing 7–2 to an Old Etonians team led by Kinnaird.[21] The following season Wanderers managed to reach the third round of the Cup, but lost again to the Etonians, after which many more key players left the club.[22]

The club struggled on into the 1880–81 season, but was forced to withdraw from the FA Cup after being unable to raise a team for the scheduled first round match. After 1881, the club was reduced to playing only one match per year, against Harrow School each Christmas.[23] A book published by the newspaper *The Sportsman* claimed that the club folded in 1884,[24] however a match at Harrow was reported in *The Times* in December 1887, which Harrow won 3–1.[25]

Reformation (2009–present)

In 2009, over 120 years after the last known Wanderers match, a "reformed" Wanderers club was founded in London, with the approval of descendants of those involved with the original club, with the intention to play occasional matches for charity.[26] [27] In 2011 the club joined the Surrey South Eastern Combination.

Wanderers team that took to the field against
Royal Engineers in May 2010

Colours and crest

2011–12 home kit

1872: FA Cup Finals

1859: Forest FC [28]

Wanderers are known to have played in orange, violet and black for at least part of their existence, although as no photographs of the team exist, the exact design is not known. A replica shirt sold in the modern era has the three colours in horizontal stripes,[29] a likely arrangement given that horizontally-striped shirts were very common during the Victorian era.[30] The programme for the club's 1875 away match with Queen's Park, however, lists Wanderers as playing in white shirts. In the absence of shirt numbering, which would not be introduced for another sixty years, the programme identifies the individual players by the colours of their stockings (socks) or caps, with Alcock and Kinnaird both listed as wearing blue and white caps and Jarvis Kenrick identified by his cerise and French grey cap, the colours of his former club Clapham Rovers.[18] The modern club's crest is derived from the Harrow School crest, which may have adorned the shirts of the original team.[31]

Grounds

Wanderers played firstly at Snaresbrook near Epping Forest, reputedly close to one of two orphanages in the area; the Infant Orphan Asylum in Snaresbrook or the Merchant Seaman's Orphan Asylum in nearby Wanstead.[32] Wanderers played at several locations between 1859 and 1887 throughout London, but predominantly at Kennington Oval (151 games), Vincent Square (31), Harrow (23), Harrow School (15) and Clapham Common (12).[33] Battersea Park has been erroneously attributed as Wanderers home ground,[34] however Wanderers played just 10 games there between 1864 and 1867.[33]

Rivalries

Prior to the formation of The Football Association in 1863, individual schools played football according to their own particular rules.[35] Due to the connection Wanderers had with Harrow School, the school's football team played Wanderers frequently – 33 games between the two were recorded between 1865 and 1883. Among the club's other regular opponents were Royal Engineers, Clapham Rovers and Civil Service.[33]

Players

A total of fifteen players who listed Wanderers as their primary club played for the England national team in international matches, as follows:[36] [37]

- C. W. Alcock (1 cap)
- Francis Birley (1 cap)
- Alexander Bonsor (2 caps)
- Frederick Green (1 cap)
- Francis Heron (1 cap)
- Hubert Heron (3 caps) – the first brothers to play for England
- Leonard Howell (1 cap)
- William Kenyon-Slaney (1 cap)
- Robert Kingsford (1 cap)
- William Lindsay (1 cap)
- Alfred Stratford (1 cap)
- Henry Wace (3 caps)
- Reginald de Courtenay Welch (1 cap)
- Charles Wollaston (4 caps)
- John Wylie (1 cap)

The following players earned international selection whilst playing at other clubs, but held membership of Wanderers:[38]

- Alexander Morten (1 cap)
- Edward Hagarty Parry (3 caps)
- John Frederick Peel Rawlinson (1 cap)
- Francis Sparks (3 caps)

Additionally, A. F. Kinnaird made one appearance for Scotland and John Hawley Edwards played his one game for Wales while registered as a Wanderers player.[39] Edwards was the first treasurer of the Welsh Football Association and one of only two players to play for England and Wales at full international level.[40] A number of Wanderers players appeared in the England vs Scotland representative matches which took place prior to what is now recognised as the first official international match.[41]

Club officials

The first Wanderers secretary was A. W. Mackenzie (1859–1864). He was succeeded by C. W. Alcock (1864–1875), Jarvis Kenrick (1875–1879) and Charles Wollaston (1879–1883). The secretary of the revived club is Mark Wilson, who has held the position since the club was set up.[42]

Records and statistics

Complete FA Cup record

Season	Performance[43] [44]
1871–72	Winners
1872–73	Winners
1873–74	Lost in quarter-finals
1874–75	Lost in quarter-finals
1875–76	Winners
1876–77	Winners
1877–78	Winners
1878–79	Lost in first round
1879–80	Lost in third round
1880–81	Withdrew
1881–82	Withdrew

Although records are incomplete, C. W. Alcock is believed to have played the most matches for the Wanderers, with at least 199 appearances, and to have scored the most goals, with at least 82.[45] He also recorded the highest goalscoring total for an individual season, with 17 known goals in the 1870–71 season, including four in a 6–1 win over Civil Service.[46] R. K. Kingsford bettered that feat when he scored five goals against Farningham in 1874, the most goals scored by a Wanderers player in a single match. The 16–0 margin of victory in the Farningham match was by far the largest win achieved by Wanderers, with no other scores in double-figures recorded.[47] The most goals conceded by Wanderers was eight, in an 8–2 defeat to Clapham Rovers in 1879; the club also lost by a six-goal margin on at least one other occasion, a 6–0 defeat to Queen's Park in 1876.[41]

Wanderers' total of five FA Cup final wins remained a record until Aston Villa won the competition for the sixth time in 1920. As of 2011, only eight clubs have won the tournament more times than the Wanderers.[48]

Honours

The club won the first FA Cup, won three in a row and appeared in the FA Cup Final five times, winning each time.[49] Wanderers hold the joint record for most consecutive wins with Blackburn Rovers and A. F. Kinnaird holds the record for appearances in a Final with nine. Wanderers are tied for ninth place with Everton, Manchester City and West Bromwich Albion for the most FA Cup wins.

- **FA Cup: Five**

  - 1871–72, 1872–73, 1875–76, 1876–77, 1877–78

References

General

- Cavallini, Rob (2005). *The Wanderers F.C.: Five Times F.A. Cup Winners.* Dog N Duck Publications. ISBN 0-9550496-0-1.
- Warsop, Keith (2004). *The Early FA Cup Finals and the Southern Amateurs.* SoccerData. ISBN 1-8994-6878-1.

Specific

[1] http://www.originalwanderers.com/

The second FA Cup trophy, pictured here, is identical in design to that won by Wanderers. The original trophy was stolen in 1895 and never recovered.

[2] Harvey, Adrian (2005). *Football: the first hundred years : the untold story.* Routledge. p. 127. ISBN 0-4153-5019-0.
[3] Alcock, C. W. (1906). *The Book of Football.* p. 255.
[4] Cavallini, p. 14
[5] Cavallini, p. 15
[6] Cavallini, p. 16
[7] Cavallini, p. 17
[8] Cavallini, p. 19
[9] Mason, Nicholas (1975). *Football!: The story of all the world's football games.* Drake Publishers. p. 24. ISBN 0-8473-1024-8.
[10] Cavallini, p. 22
[11] Cavallini, p. 25
[12] Cavallini, pp. 120–123
[13] Cavallini, p. 29
[14] Cavallini, p. 126
[15] Warsop, p. 40
[16] Warsop, p. 41
[17] Cavallini, p. 43
[18] Soar, Phil; Martin Tyler (1983). *Encyclopedia of British Football.* Willow Books. p. 64. ISBN 0-0021-8049-9.
[19] Cavallini, p. 47
[20] Cavallini, p. 51
[21] Cavallini, p. 138
[22] Cavallini, p. 140
[23] Cavallini, p. 56
[24] Pelham Warner, ed. (1917). *British Sports and Sportsmen: Cricket and Football.* J. G. Hammond & Co. Ltd.. p. 237.
[25] "Football". *The Times.* 21 December 1887.
[26] "Wanderers website" (http://www.pitchero.com/clubs/wanderersfc/a/reformation-21500.html). Wanderers F.C.. . Retrieved 21 August 2011.
[27] "Alcock family unveil plaque for 'Father of Sport'" (http://www.journallive.co.uk/north-east-news/todays-news/2010/11/20/plaque-honours-north-creator-of-football-history-61634-27683367/). Journal Live. . Retrieved 21 August 2011.
[28] "Historical Football Kits" (http://www.historicalkits.co.uk/Articles/Olde_Curiosity_Shoppe.htm). Historical Football Kits. . Retrieved 20 August 2011.
[29] "Wanderers 1872 FA Cup Winners" (http://www.toffs.com/Wanderers-1892-FA-Cup-Winners/productinfo/1237/). The Old Fashioned Football Shirt Company. . Retrieved 14 February 2011.

[30] Warsop, p. 143

[31] "Brief History of Football Kit Design" (http://www.historicalkits.co.uk/Articles/History.htm). Historical Football kits. . Retrieved 20 August 2011.

[32] Cavallini, p. 14

[33] Cavallini, pp. 116–141

[34] Blythe Smart, J. (2003). *The Wow Factor*. Blythe Smart Publications. p. 144.

[35] Greenland, W. E. (1965). *The History of the Amateur Football Alliance*. Standard Printing & Publishing Company. p. 8.

[36] Cavallini, p. 142

[37] "Wanderers" (http://www.englandfootballonline.com/TeamClubs/Clubs/Wanderers.html). *England Players' Club Affiliations*. Englandfootballonline. . Retrieved 17 February 2011.

[38] Cavallini, pp. 104–114.

[39] Cavallini, p. 102

[40] Football Online "England Football Online" (http://www.englandfootballonline.com/TeamHist/TrivPlyrsOtherNatTeams.html=England). Football Online. Retrieved 20 August 2011.

[41] Cavallini, p. 137

[42] "Officials" (http://www.pitchero.com/clubs/wanderersfc/officials/). Wanderers F.C.. . Retrieved 21 August 2011.

[43] "The FA Cup Archive" (http://www.thefa.com/TheFACup/FACompetitions/TheFACup/Archive). The Football Association. . Retrieved 14 February 2011.

[44] Collett, Mike (2003). *The Complete Record of the FA Cup*. Sports Books. p. 630. ISBN 1-8998-0719-5.

[45] Cavallini, p. 65

[46] Cavallini, p. 126

[47] Cavallini, p. 129

[48] Barnes, Stuart, ed. (2010). *The Nationwide Football Annual 2010–2011*. Sportsbooks. p. 162. ISBN 1-8998-0793-4.

[49] "Cup Final Results" (http://www.thefa.com/TheFACup/FACompetitions/TheFACup/History/CupFinalResults). The Football Association. . Retrieved 22 January 2011.

External links

- Wanderers (http://www.fchd.info/WANDERER.HTM) at the Football Club History Database
- Official website (http://http://www.originalwanderers.com/)

Daventry_District

<table>
<tr><td colspan="2" align="center">Daventry District</td></tr>
<tr><td colspan="2" align="center">— District —</td></tr>
<tr><td colspan="2" align="center">Shown within Northamptonshire</td></tr>
<tr><td>Sovereign state</td><td>United Kingdom</td></tr>
<tr><td>Constituent country</td><td>England</td></tr>
<tr><td>Region</td><td>East Midlands</td></tr>
<tr><td>Administrative county</td><td>Northamptonshire</td></tr>
<tr><td>Founded</td><td></td></tr>
<tr><td>Admin. HQ</td><td>Daventry</td></tr>
<tr><td>Government</td><td></td></tr>
<tr><td>• Type</td><td>Daventry District Council</td></tr>
<tr><td>• Leadership:</td><td>Alternative - Sec.31</td></tr>
<tr><td>• Executive:</td><td>Conservative</td></tr>
<tr><td>• MP:</td><td>Chris Heaton-Harris</td></tr>
<tr><td>Area</td><td></td></tr>
<tr><td>• Total</td><td>unknown operator: u'strong' sq mi (662.6 km^2)</td></tr>
<tr><td>Area rank</td><td>58th</td></tr>
<tr><td>Population (2010 est.)</td><td></td></tr>
<tr><td>• Total</td><td>79,000</td></tr>
<tr><td>• Rank</td><td>Ranked 288th</td></tr>
<tr><td>Time zone</td><td>Greenwich Mean Time (UTC+0)</td></tr>
<tr><td>• Summer (DST)</td><td>British Summer Time (UTC+1)</td></tr>
<tr><td>Postcode</td><td></td></tr>
<tr><td>ISO 3166-2</td><td></td></tr>
<tr><td>ONS code</td><td>34UC</td></tr>
</table>

OS grid reference	
NUTS 3	
Ethnicity	98.0% White
Website	daventrydc.gov.uk [1]

The Daventry district **also known as Chaventry** is the largest local government district of western Northamptonshire, England. The district is named after the town of Daventry which is the administrative headquarters and largest town. The district is predominantly rural, other significant settlements include Brixworth, Long Buckby and Weedon Bec.

The Daventry district was created on 1 April 1974, under the Local Government Act 1972, by a merger of the historic municipal borough of Daventry, with the Daventry Rural District and most of the Brixworth Rural District. The town of Daventry became an unparished area with Charter Trustees and remained so until 2003 when a civil parish was created, roughly corresponding with the boundaries of the former borough, so allowing Daventry to have its own town council.[2]

Settlements in Daventry district

- Althorp, Arthingworth, Ashby St Ledgers
- Badby, Barby, Boughton, Braunston, Brington, Brixworth, Brockhall, Byfield
- Canons Ashby, Chapel Brampton, Charwelton, Church Brampton, Church Stowe, Clay Coton Clipston, Cold Ashby, Coton, Cottesbrooke, Creaton, Crick
- Daventry, Dodford, Draughton
- East Farndon, East Haddon, Elkington, Everdon
- Farthingstone, Fawsley, Flore
- Great Brington, Great Oxendon, Guilsborough
- Hanging Houghton, Hannington, Harlestone, Haselbech, Hellidon, Holcot, Holdenby, Hollowell
- Kelmarsh, Kilsby
- Lamport, Lilbourne, Little Brington, Long Buckby, Lower Catesby
- Maidwell, Marston Trussell, Moulton
- Naseby, Newnham, Norton
- Old, Overstone
- Pitsford, Preston Capes
- Ravensthorpe
- Scaldwell, Sibbertoft, Spratton, Stanford-on-Avon, Staverton, Sulby
- Teeton, Thornby
- Upper Catesby, Upper Stowe
- Walgrave, Watford, Weedon Bec, Welford, Welton, West Haddon, Whilton, Winwick, Woodford Halse
- Yelvertoft

Historical settlements

- Bannaventa

Energy policy

In May 2006, a report commissioned by British Gas[3] showed that housing in the district of Daventy produced the 7th highest average carbon emissions in the country at 7,276 kg of carbon dioxide per dwelling.

References

[1] http://www.daventrydc.gov.uk/

[2] Daventry Town Council (http://www.daventrytowncouncil.gov.uk)

[3] (http://www.britishgasnews.co.uk/managed_content/files/pdf/greenCity.pdf) *britishgasnews.co.uk*

Henry_Holmes_Stewart

Rev. **Henry Holmes Stewart** (8 November 1847 – 20 March 1937) was a Scottish clergyman who was a member of the Wanderers team which won the FA Cup in 1873. He also played for the Scottish team in 1872 in the last of the series of representative football matches against England.

Family and education

Stewart was born in Cairnsmuir, near Newton Stewart, Kirkcudbrightshire, the son of James Stewart[1] and Elizabeth MacLeod.[2] His brothers included James (1840–1938)[3] and Ravenscroft (1845–1921),[4] both of whom also attended Trinity College.

He attended Repton School and Loretto College, Edinburgh before going up to Trinity College, Cambridge in 1867.[2] He graduated in 1871 with a B.A. and was awarded his M.A. in 1874.[2]

On 28 July 1874, he married Lady Beatrice Diana Cecilia Carnegie,[5] daughter of James Carnegie, 9th Earl of Southesk and Lady Catherine Hamilton Noel.[1]

Cricket career

At Repton School, he was an outstanding cricketer and was in the school team from 1865 to 1867; in his final season, he was the school's best batsman.[6] He also played cricket for Cambridge University although he did not play any first class matches.[6] He also played for M.C.C. and I Zingari.[7]

Following his move to Glamorgan, he played village cricket, continuing well into the twentieth century.[6]

Football career

After leaving university, he joined the Wanderers club. He made his debut for them on 4 March 1872 at Kennington Oval in the semi-final of the FA Cup against the Scottish team, Queens Park;[8] this was the first time that a Scottish side had visited London and the Scots' travelling expenses were met by public subscription in Glasgow.[9] The match ended in a 0–0 draw; as Queens Park were unable to raise the cost of a second trip to London, they withdrew from the competition, leaving Wanderers to go through to the final.[9]

Two weeks before the FA Cup semi-final, Stewart was a member of the Scottish team that played England in what was to be the last of the series of representative matches between the two countries. The match ended with a 1–0 victory for the English.[10] [11] In a match report, Stewart was praised for his "untiring forward play throughout".[12]

In the next season, Stewart played frequently for the Wanderers making eight appearances. He was variously described as "keeps well on the ball and never flags"[8] and "sticks close to the ball and follows up hard; a very useful forward".[6] As holders, Wanderers were given a "bye" to the Cup Final in which Stewart was selected as one of the eight forwards. The final, played against Oxford University at Lillie Bridge on 29 March 1873 ended in a 2–0 victory for the Wanderers.[13]

Stewart played three more matches for the Wanderers in 1873–74 before his clerical career took him away from London.[8]

Clerical career

Stewart was ordained as a deacon in London in 1872 and as a priest in 1873. He was curate at St. John's, Holborn from 1872 to 1874 and then vicar at East Witton, North Riding of Yorkshire from 1874 to 1878. He was then rector at Brington, Northamptonshire (1878–1898), vicar at Porthkerry with Barry, Glamorgan (1898–1914), vicar at St. Lythan's, Glamorgan (1914–1925) and, finally, rector at Michaelston-le-Pit, Glamorgan from 1925 to 1935.[2]

He died on 20 March 1937, aged 89 years, at his home at Dinas Powys, Glamorgan.[6]

References

[1] "Reverend Henry Holmes Stewart" (http://thepeerage.com/p29551.htm#i295505). thepeerage.com. 10 June 2008. . Retrieved 9 April 2011.

[2] Venn, J.; Venn, J. A., eds. (1922–1958). " Stewart, Henry (http://venn.lib.cam.ac.uk/cgi-bin/search.pl?sur=&suro=c&fir=&firo=c& cit=&cito=c&c=all&tex=STWT866HH&sye=&eye=&col=all&maxcount=50)". *Alumni Cantabrigienses (10 vols)* (online ed.). Cambridge University Press.

[3] Venn, J.; Venn, J. A., eds. (1922–1958). " Stewart, James (http://venn.lib.cam.ac.uk/cgi-bin/search.pl?sur=&suro=c&fir=&firo=c& cit=&cito=c&c=all&tex=STWT859J&sye=&eye=&col=all&maxcount=50)". *Alumni Cantabrigienses (10 vols)* (online ed.). Cambridge University Press.

[4] Venn, J.; Venn, J. A., eds. (1922–1958). " Stewart, Ravenscroft (http://venn.lib.cam.ac.uk/cgi-bin/search.pl?sur=&suro=c&fir=& firo=c&cit=&cito=c&c=all&tex=STWT864R&sye=&eye=&col=all&maxcount=50)". *Alumni Cantabrigienses (10 vols)* (online ed.). Cambridge University Press.

[5] "Lady Beatrice Diana Cecilia Carnegie" (http://thepeerage.com/p5810.htm#i58100). thepeerage.com. 10 June 2008. . Retrieved 9 April 2011.

[6] Warsop, Keith (2004). *The Early FA Cup Finals and the Southern Amateurs*. SoccerData. pp. 126–127. ISBN 1-899468-78-1.

[7] "Teams Henry Stewart played for" (http://cricketarchive.com/Archive/Players/421/421573/all_teams.html). cricketarchive. . Retrieved 10 April 2011.

[8] Cavallini, Rob (2005). *The Wanderers F.C. – "Five times F.A. Cup winners"*. Dog N Duck Publications. pp. 99–100. ISBN 0-9550496-0-1.

[9] Cavallini, Rob. *The Wanderers F.C. – "Five times F.A. Cup winners"*. pp. 36–37.

[10] "England 1 Scotland 0" (http://www.englandfootballonline.com/Seas1872-00/1871-72/M00UOSco1872.html). *England Unofficial Matches*. englandfootballonline. 24 February 1872. . Retrieved 10 April 2011.

[11] "England 1 Scotland 0" (http://www.londonhearts.com/scotland/games/18720224.html). London Hearts. 24 February 1872. . Retrieved 10 April 2011.

[12] "England 1 Scotland 0" (http://www.londonhearts.com/scores/images/1872/1872022403.htm). *Scottish football reports*. London Hearts. 24 February 1872. . Retrieved 10 April 2011.

[13] Warsop. *The Early FA Cup Finals and the Southern Amateurs*.

External links

- Cricket career details on Cricket Archive (http://cricketarchive.com/Archive/Players/421/421573/421573.html)

Nobottle

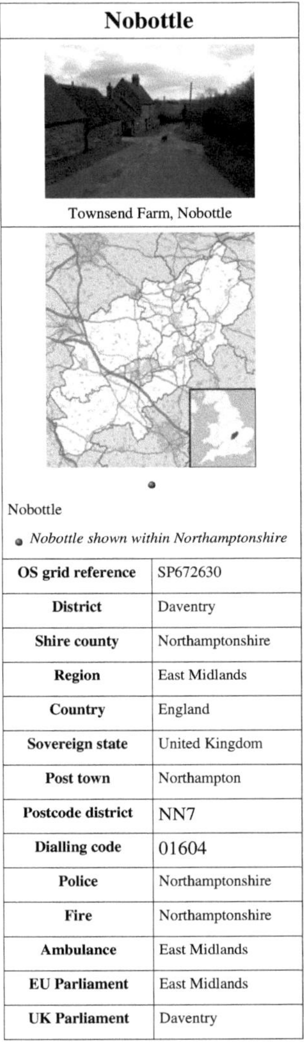

Nobottle

Townsend Farm, Nobottle

Nobottle

Nobottle shown within Northamptonshire

OS grid reference	SP672630
District	Daventry
Shire county	Northamptonshire
Region	East Midlands
Country	England
Sovereign state	United Kingdom
Post town	Northampton
Postcode district	NN7
Dialling code	01604
Police	Northamptonshire
Fire	Northamptonshire
Ambulance	East Midlands
EU Parliament	East Midlands
UK Parliament	Daventry

Further information: List of lost settlements in Northamptonshire

Nobottle is a hamlet in the Daventry district of the county of Northamptonshire in England. It borders the Althorp estate, which owns much of the property. Nobottle used to have a 600yd rifle range (the only one in Northamptonshire), now shut by the MOD some 20 years. The Midshires Way long distance footpath passes through Nobottle.

External links

- Map sources for Nobottle

Rector

The word **rector** ("ruler", from the Latin *regere* and *rector* meaning "ruler" in Latin) has a number of different meanings; it is widely used to refer to an academic, religious or political administrator.

The term and office of a rector are called a *rectorate*.

Academic rectors

The rector is the highest academic official of many universities and in certain other institutions of higher education, as well as even in some secondary-level schools.

Inauguration of Rector Lubomír Dvořák (Palacký University)

The title is used widely in universities across Europe.[1] It is also very common in Latin American countries.[2] It is also used in Russia, Pakistan, the Philippines, Indonesia and Israel, all of which are strongly influenced by European traditions. In some universities, the title is phrased in an even loftier manner, as *Rector Magnificus* or *Lord Rector*.

A notable exception to this terminology is in England and elsewhere in Great Britain, where the head of a university has traditionally been referred to as a "Chancellor". This pattern has been followed in the Commonwealth, the United States, and other countries under British influence. In Scotland, many universities are headed by a Chancellor, with the Lord Rector designated as an elected representative of students at the head of the university court.

Academic rectors in Europe

The head of a university in Germany is called a *president, rector magnificus* (men) or *rectrix magnifica* (women), as in some Belgian universities (notably the oldest, Katholieke Universiteit Leuven). In Dutch universities, the *rector magnificus* is the most publicly prominent member of the board, responsible for the scientific agenda of the university. The rector is not the chair of the board. The chair has, in practice, the most influence over the management of the University.

In some countries, including Germany, the position of head teacher in secondary schools is also designated as rector. In the Netherlands, the terms Rector and Conrector (assistant head) are used commonly for high school directors. This is also the case in some Maltese secondary schools.

In the Scandinavian countries, the head of a university or a gymnasium (higher secondary schools) is called a *rektor*. In Sweden and Norway, this term is also used for the heads of primary schools. In Finland, the head of a primary school is called a rector (*rehtori*) provided the school is of sufficient size in terms of faculty and students. Otherwise the title is headmaster (*koulunjohtaja*).

In the Iberian Peninsula, Portugal's and Spain's university heads or presidents have the title *Magnífico Reitor/Rector Magnífico*, and are usually styled, in official ceremonies, with the denomination of "Most Excellent and Illustrious Sir or Lord". For example, in Portugal, the rector of the University of Coimbra, the oldest Portuguese university, is referred to as *Magnífico Reitor Professor Doutor (Rector's name)* ("Rector Magnificus Professor Doctor (Rector's Name)"). In Spain, the Rector of the University of Salamanca, the oldest on the Iberian Peninsula, is usually styled according to academic protocol as *Excelentísimo e Ilustrísimo Señor Profesor Doctor Don (Rector's name), Rector Magnífico de la Universidad de Salamanca* ("The Most Excellent and Most Illustrious Lord Professor Doctor Don

(Rector's name), *Rector Magnificus* of the University of Salamanca").

Czech Republic

The heads of Czech Universities are called the *rektor*. The Rector acts in the name of the University and decides the University's affairs unless prohibited by law. The Rector is nominated by the University Academical Senate and appointed by the President of the Czech Republic. The nomination must be agreed by a simple majority of all Senators, while a dismissal must be agreed by at least three fifths of all Senators. The vote to elect or repeal a Rector is secret. The term of office is four years and a person may hold it for at most two consecutive terms.

The Rector appoints vice-rectors (*pro-rektor*), who act as deputies to the extent the Rector determines. Rectors' salaries are determined directly by the Minister of Education.

Among the most important rectors of Czech Universities were reformer Jan Hus, physician Jan Jesenius and representative of Enlightenment Josef Vratislav Monse. The first female rector became in 1950 Jiřina Popelová (Palacký University of Olomouc).

The Rectors are addressed *"Your Magnificence Mister Rector"* (*"Vaše Magnificence pane rektore"*).

Denmark

In Danish, *rektor* is the title used in referring to the heads of universities, gymnasiums, schools of commerce and construction, etc. Generally *rektor* may be used for the head of any educational institution above the primary school level, where the head is commonly referred to as a 'skoleinspektør' (Headmaster; Inspector of the school). In universities, the second-ranked official of governance is known as *prorektor*.

England

At Oxford and Cambridge, English universities which are formally headed by *chancellors*, most colleges are headed by a *master* or a *principal* as the chief academic. In a few colleges, the person filling this role is called a *president* or a *warden*. At two of the Oxford colleges, Lincoln College and Exeter College, the head is called a "rector." The leader of all of the organization is called a "chancellor" in both Oxford University and Cambridge University.[3] [4]

At Durham University the University is formally headed by the Chancellor, but due to its ecclesiastical background, the *formal* head of St Chad's College is the *Rector*, while the head is the *Principal* or *President*.

The University of London has a Chancellor (a ceremonial post) and a Vice-Chancellor (equivalent to a managing director). All colleges have a chief academic as head, using a variety of titles. At University College London, the head is the *Provost*; at King's College London the head is the *Principal*; at Imperial College London the head is the *Rector*; and the London School of Economics is headed by a *Director*.

At most other universities in England, the Chancellor is the ceremonial head whilst the Vice-Chancellor is the chief academic. The Vice-Chancellor of Liverpool Hope University also takes the role of Rector.

Germany

The chief academic in most german universities is called "Rektor" (rector), in less traditional universities he is called "Präsident" (president).

Iceland

The *rektor* is term used for the headmaster or headmistress of Icelandic universities and of some gymnasia.

Italy

In Italy the rector (*Magnifico Rettore*) is the head of the university and *Rappresentante Legale* of the university. He or she is elected by an electoral body composed of all *Professori ordinari and Associati* (full and associate professors), the two highest ranks of the Italian university faculty, and of representatives of the *Ricercatori* (lowest rank of faculty), and of the staff.

The term of a *rettore* is usually four to five years, in accordance with the *statuto* (constitution of the university). The Rettore is styled and formally greeted as *Magnifico Rettore*.

Netherlands

In the Netherlands, the rector is the principal of a high school. The rector is supported by conrectors (deputy rectors who can take his place).

In Dutch universities, the *Rector Magnificus* is the member of the executive board of the university responsible for the scientific vision and quality of the university. The *rector magnificus* is a full professor. The ceremonial responsibilities of the *rector magnificus* are to open the academic year, and to preside over PhD defenses. In practice of the latter function, the rector is usually replaced by a member of the PhD examination board of the university, which consists solely of full professors.

Norway

A *rektor* is the headmaster of a primary school, secondary school, private school, high school, college or university.

Portugal

In Portugal, the Rector (Portuguese: *Reitor*) is the highest official of each university. The official complete title is *Magnífico Reitor* (Magnificent Rector). Each university faculty is headed by a director or a president of the directorate council, and the rector heads all of them.

Until 1974, the director of each Lyceum (high school) also had the title of Rector.

Scotland

In Scotland, the position of Rector exists in the four Ancient Universities - (St Andrews, Glasgow, Aberdeen and Edinburgh) (in order of foundation) and at Dundee, which is considered to have Ancient status as a result of its early connections to St. Andrews University.

The post (officially *Lord Rector*, but by normal usage just *Rector*) was made an integral part of these universities by the Universities (Scotland) Act 1889. The nominal head of an ancient university in Scotland is its Chancellor with the day-to-day functions of the chief operating officer vested in the vice-chancellor, who also holds the title of Principal and is referred to as the *Principal Vice-Chancellor*. The Rector is the third-ranking official of university governance and chairs meetings of the university court, the governing body of the university, and is elected at regular intervals (usually every three years to enable every undergraduate who obtains a degree to vote at least once) by their matriculated student bodies. An exception exists in Edinburgh, where the Rector is elected by both students and staff.[5]

The role of the rector is considered by many students to be integral to their ability to shape the universities' agenda, and one of the main functions of the rector is to represent the interests of the student body. To some extent the office of rector has evolved into more of a figurehead role, with a significant number of celebrities and personalities elected as rectors, such as Stephen Fry and Lorraine Kelly at Dundee, Clarissa Dickson Wright at Aberdeen, and John Cleese and Frank Muir at St. Andrews, and political figures, such as Mordechai Vanunu at Glasgow. In many cases, particularly with high-profile rectors, attendance at the university court in person is rare; however, the Rector nominates an individual (normally a member of the student body) with the title of *Rector's Assessor*, to exercise his/her functions.

The Rt. Hon. Gordon Brown, the former Prime Minister of the United Kingdom, was Rector of the University of Edinburgh while a student there, but since then most universities have amended their procedures to forbid currently matriculated students from standing for election.

The current Rector of the University of Aberdeen is Maitland Mackie, owner and chairman of Mackie's of Scotland. The current Rector of the University of Dundee is the Scottish actor Brian Cox, CBE. The current Rector of the University of Glasgow is Charles Kennedy MP, former leader of the Liberal Democrats and a former President of the Glasgow University Union. He was first elected in 2008 and was re-elected in February 2011. The current Rector of the University of St Andrews is Kevin Dunion OBE the first and current Scottish Information Commissioner.

The Head teacher of a Scottish High School/Secondary School is often known by the title of *Rector*.

Spain

In Spain, *Rector* or *Rector Magnífico* (magnific rector, from Latin *Rector Magnificus*) is the highest administrative and educational office in a university, equivalent to that of President or Chancellor of an English-speaking university, but holding all the powers of a vice-chancellor; they are thus the head of the academi in Universities. Formally styled as *"Excelentísimo e Ilustrísimo Señor Profesor Doctor Don N, Rector Magnífico de la Universidad de X"* (Most Excellent and Illustrious Lord Professor Doctor Don N, Rector Magnificus of the University of X), it is an office of high dignity within Spanish society, usually being highly respected. It is not strange to see them appear in the media, especially when some academic-related subject is being discussed and their opinion is requested.

Spanish Rectors are chosen from within the body of university full Professors (*Catedráticos* in Spanish); it is compulsory for anyone aspiring to become a rector to have been a Doctor for at least 6 years before his election, and to have achieved Professor status, holding it in the same university for which he is running. Usually, when running for the election the rector will need to have chosen the vice-rectors (*vicerrectores* in Spanish) who will occupy several sub-offices in the university. Rectors are elected directly by free and secret universal suffrage of all the members of the university, including students, lecturers, readers, researchers, and civil servants,... However, the weight of the vote in each academic sector is different: the total student vote usually represents 20% of the whole, no matter how many students there are; the votes of the entire group made up of professors and readers (members of what formerly was known as the *Claustro* (*cloister*)) usually counts for about 40-50% of the total; lecturers, researchers (including Ph.D. students and others) and non-doctoral teachers, about 20% of the total; and the remainder (usually some 5-10%) is left for non-scholarly workers (people in administration, etc.) in the university. Spanish law allows those percentages to be changed according to the situation of each university, or even not to have a direct election system. Indeed, in a few universities the Rector is chosen indirectly; the members of the modern *Claustro* (a sort of electoral college or parliament in which all the above-mentioned groups are represented) is chosen first, and then the Claustro selects the Rector.

Rectors hold their office for four years before another election is held, and there is no limit to the number of re-election terms. However, only the most charismatic and respected rectors have been able to hold their office for more than two or three terms. Of those, some have been notable Spanish scholars, such as Basque writer Miguel de Unamuno, Rector of the University of Salamanca from 1901 until 1936.

Sweden

Rektor is the title for the highest-ranked administrative and educational leader for an academic institution, such as a primary school, secondary school, private school, high school, college or university. The *rektors* of state-run colleges and universities are political appointees of the government. The adjunct of a *rektor* at a university is called a *prorektor* and is appointed by the institution's board.

In the older universities of Uppsala and Lund, the *rektor* is titled *rector magnificus* (men), or *rectrix magnifica* (women). Younger universities have in more recent years started using the Latin honorary title in formal situations, such as in honorary speeches or graduation ceremonies.

Central and Eastern Europe and Turkey

The Rector is the head of most universities and other higher educational institutions in at least parts of Central and Eastern Europe, such as Bulgaria, Croatia, Poland, Romania, Russia, Macedonia, Serbia, Slovenia, Turkey and Ukraine. The rector's deputies are known as "pro-rectors".

Academic rectors in North America

Canada

As in most Commonwealth and British-influenced countries, the term "rector" is not commonly used in Canada.

Quebec's universities, both francophone (e.g., *Université de Montréal*) and anglophone (e.g., Concordia University), use the term (*recteur* or *rectrice* in French) to designate the head of the institution. In addition, the historically French-Catholic, and currently bilingual, Saint Paul University in Ottawa, Ontario uses the term to denote its head. St. Paul's College, the Roman Catholic College of the University of Manitoba, uses the term 'rector' to designate the head of the College. St. Boniface College, the French College of the University of Manitoba, uses 'recteur' or 'rectrice' to designate the head of the College.

Queen's University (Kingston, Ontario) uses the term "rector". The term refers to a member of the student body elected to work as an equal with the Chancellor and Principal. The Badge of Office of the Rector of Queen's University was registered with the Canadian Heraldic Authority on October 15, 2004.[6]

United States

Most U.S. colleges use the titles "president" for the chief executive of the college and "chair of the board of trustees" for the head of the body that legally "owns" the college. The terms "president" and "chancellor" are used for the chief executive of some universities and university systems, depending on the institution's statutes (some state university systems have both presidents of constituent colleges and a chancellor of the overall system, or vice versa).

Several notable exceptions exist in the Commonwealth of Virginia: the University of Virginia (Charlottesville), University of Mary Washington (Fredericksburg), George Mason University (Fairfax), Virginia State University (Petersburg), Virginia Commonwealth University (Richmond), Washington and Lee University (Lexington), the College of William and Mary (Williamsburg), Old Dominion University (Norfolk), and Virginia Tech (Blacksburg) use the term "Rector" to designate the head of the Board of Visitors. The College of William and Mary also has a "Chancellor" who acts in a ceremonial capacity.

From 1701-1745, the head of the school that was to become Yale University was termed the "rector". As head of Yale College, Thomas Clap was both the last to be called "rector" (1740–1745) and the first to be referred to as president (1745–1766). Modern custom omits the use of the term "rector" and identifies Abraham Pierson as the first Yale president (1701–1707). Clap is construed to have been the fifth in the sequence of men who were Yale's leaders.[7]

Several Catholic colleges and universities, particularly those run by religious orders of priests (such as the Jesuits) formerly employed the term "rector" to refer to the school's chief officer. In many cases, the rector was also the head of the community of priests assigned to the school, so the two posts – head of the university and local superior of the priests – were merged in the role of rector (*See "Ecclesiastical rectors" below*). This practice is no longer followed, as the details of the governance of most of these schools have changed.

Academic rectors in Australia

The term "rector" is uncommon in Australian academic institutions. The executive head of an Australian university has traditionally been given the British title Vice-Chancellor, although in recent times the American term President has also been adopted. The term rector is used by some academic institutions, such as the University of Melbourne residential college, Newman College; the private boys' school, Xavier College; and the University of Sydney residential college, St John's College (Benedictine).

The title Rector is sometimes used for the head of a subordinate and geographically separate campus of a university. For example, the executive head of the Australian Defence Force Academy in Canberra, which is a campus of the University of New South Wales in Sydney is a Rector, as is the head of the Cairns campus of James Cook University, based at Townsville.

Academic rectors in New Zealand

The title is used in New Zealand for the Headmaster of some independent schools, such as Lindisfarne College, and a number of state schools for boys, including Otago Boys' High School, King's High School, Dunedin, Waitaki Boys' High School, Timaru Boys' High School, Palmerston North Boys' High School and Southland Boys' High School showing the Scots' involvement in the foundation of those schools.

Academic rectors in Asia

India

The heads of certain Indian boarding schools are called Rectors. The head or principal of a Catholic schools in India is also called a rector.

Japan

During the years of the Tokugawa shogunate (1601–1868), the rector of Edo's Confucian Academy, the *Shōhei-kō* (afterwards known at the *Yushima Seidō*), was known by the honorific title *Daigaku-no kami* which, in the context of the Tokugawa hierarchy, can effectively be translated as "Head of the State University". The rector of the *Yushima Seidō* stood at the apex of the country-wide educational and training system which was created and maintained with the personal involvement of successive shoguns. The position as rector of the *Yushima Seidō* became hereditary in the Hayashi family.[8] The rectors' scholarly reputation was burnished by the publication in 1657 of the seven volumes of *Survey of the Sovereigns of Japan* (*Nihon Ōdai Ichiran*)[9] and by the publication in 1670 of the 310 volumes of *The Comprehensive History of Japan* (*Honchō-tsugan*).[10]

Colonnade at the reconstructed *Yushima Seidō* in Tokyo. The hereditary rectors of this Edo period institution were selected from the Hayashi clan.

Malaysia

In this Commonwealth nation, the term *Rektor* is used to refer to the highest administrative official in several universities and higher education institutions in Malaysia, such as the International Islamic University Malaysia in Gombak and the Universiti Teknologi MARA in Perak. A *Rektor* is comparable to the position of *Naib Canselor*, or Vice-Chancellor, in other higher education institutions, as the *Rektor* answers to the *Canselor*.

Myanmar

The term *Rector* (Burmese:) is used to refer to the highest official of universities in Myanmar. Each university department is headed by a professor, who is responsible to the rector. Nowadays, given the large dimensions of some universities, the position of pro-rector has emerged, just below that of the rector. Pro-rectors are in charge of managing particular areas of the university, such as research or undergraduate education.

Philippines

The term *Rector* or *Rector Magnificus* is used to refer to the highest official in prominent Catholic universities and colleges such as the University of Santo Tomas, Colegio de San Juan de Letran and San Beda College. The rector typically sits as chair of the university board of trustees. He exercises policy-making as well as general academic, managerial, and religious functions over all university academic and non-academic staff.

In the University of Santo Tomas, the highest individual academic award conferred on a graduating college student is the *Rector's Award for Academic Excellence.*

Rev. Fr. Anscar J. Chupungco, OSB, a world-renowned liturgist and theologian, served as the twentieth rector-president of San Beda College. Prior to this, he was rector-magnificus of the Pontifical Liturgical Institute and the Pontifical Ateneo San Anselmo, both in Rome.

Thailand

The term *Rector* is not widely used to refer to the highest executive position in Thai universities (Thai:อธิการบดี; RTGS: Athikan Bodi), compared to the term *President.* Thammasat University adopts this term for this position to reflect its tradition associated with the French education system where Pridi Banomyong, Thammasat's founding father was educated.

Academic rectors in South America

Argentina

The term *Rector* is used to refer to the highest official of universities, and university-owned high schools (e.g. Escuela Superior de Comercio Carlos Pellegrini) in Argentina. Each school (Spanish:*Facultad*) has its own rector, acting as school director.

Brazil

The term *Rector* (Portuguese: *Reitor*) is used to refer to the highest official of universities in Brazil. Each faculty is headed by a director, who is under the authority of the rector. Nowadays, given the large size of some universities, the position of pro-rector has emerged below that of the rector. The pro-rector is in charge of managing a particular area of the university, such as research or undergraduate education.

Ecclesiastical rectors

In ancient times bishops, as rulers of cities and provinces, especially in the Papal States, were called rectors, as were administrators of the patrimony of the Church (e.g. *rector Siciliæ*). The term 'Rector' was used by Pope Gregory the Great in the "Regula Pastoralis" as equivalent to pastor.

Roman Catholic hierarchies

In the Roman Catholic Church, a rector is a person who holds the office of presiding over an ecclesiastical institution. This institution might be a particular building—like a church or shrine—or it could also be an organization, such as a parish, a mission or quasi-parish, a seminary or house of studies, a university, a hospital, or a community of clerics or religious.

The Canon law of the Catholic Church explicitly mentions as special cases three offices of rectors: rectors of seminaries (c. 239 & c. 833 #6); rectors of churches that do not belong to a parish, a chapter of canons, or a religious order (c. 556–553); and rectors of Catholic universities (c. 443 §3 #3 & c. 833 #7). However, these are not the only officials who exercise their functions using the title of rector.

Since the term rector refers to the function of the particular office, a number of officials are not referred to as rectors even though they are rectors in actual practice. The diocesan bishop, for instance, is himself a rector, since he presides over both an ecclesiastical organization (the diocese) and an ecclesiastical building (his cathedral). In many dioceses, the bishop delegates the day-to-day operation of the cathedral to a priest, who is often incorrectly called a rector but whose specific title is *plebanus* or "people's pastor", especially if the cathedral is also a parish. Therefore, because a priest is assigned as head of a cathedral parish, he cannot be both rector and pastor, as a rector cannot canonically hold title over a parish (c.556). As a further example, the pastor of a parish (*parochus* in Latin) is pastor (not rector) over both his parish and the parish church. Finally, a president of a Catholic university is rector over the university and, if a priest, often the rector of any church that the university may operate on the basis that it is not a canonical establishment of a parish (c. 557 §3).

In some religious congregations of priests, rector is the title of the local superior of a house or community of the order (for instance, a community of several dozen Jesuit priests might include the pastor and priests assigned to a parish church next door, the faculty of a Jesuit high school across the street, and the priests in an administrative office down the block, but the community as a local installation of Jesuit priests is headed by a rector).

Rector general is the title given to the superior general of certain religious orders, e.g. the Clerics Regular of the Mother of God, Pallottines.

There are some other uses of this title, such as for residence hall directors at the University of Notre Dame which were once (and to some extent still are) run in a seminary-like fashion. This title is used similarly at the University of Portland, another institution of the Congregation of Holy Cross.

The pope himself has been called 'rector of the world', in the (now discontinued) ceremony of the conferring of the papal tiara as part of his formal installation after election.

A now obsolete use of the term existed in the United States prior to the formulation of the 1917 Code of Canon Law. Canon Law grants a type of tenure to pastors (*parochus*) of parishes, giving them certain rights against arbitrary removal by the bishop of their diocese. In order to preserve their flexibility and authority in assigning priests to parishes, bishops in the United States until that time did not actually appoint priests as pastors, but as "permanent rectors" of their parishes: the "permanent" gave the priest a degree of confidence in the security in his assignment, but the "rector" rather than "pastor" preserved the bishop's absolute authority to reassign clergy. Hence, many older parishes list among their early leaders priests with the postnominal letters "P.R." (as in, a plaque listing all of the pastors of a parish, with "Rev. John Smith, P.R."). This practice was discontinued and today priests are normally assigned as pastors of parishes, and bishops in practice reassign them at will (though there are still questions about the canonical legality of this).

Anglican churches

In Anglican churches, a *rector* is one type of parish priest. Historically, parish priests in the Church of England were divided into rectors, *vicars*, and *perpetual curates*. The parish clergy and church was supported by tithes—like a local tax (traditionally, as the etymology of *tithe* suggests, of ten percent) levied on the personal as well as agricultural output of the parish. Roughly speaking, the distinction was that a rector directly received both the greater and lesser tithes of his parish while a vicar received only the lesser tithes (the greater tithes going to the lay holder, or *impropriator*, of the living); a perpetual curate with a small cure and often aged or infirm received neither greater nor lesser tithes, and received only a small salary (paid sometimes by the diocese). Quite commonly, parishes that had a rector as priest also had glebe lands attached to the parish. The rector was then responsible for the repair of the chancel of his church—the part dedicated to the sacred offices—while the rest of the building was the responsibility of the parish. This rectorial responsibility persists, in perpetuity, with the occupiers of the original rectorial land where it has been sold. This is called chancel repair liability, and affects institutional, corporate and private owners of land once owned by around 5,200 churches in England and Wales.[11] (See also Church of England#Organisation) Today, the roles of a rector and a vicar are essentially the same. Which of the two titles is held by the parish priest is historical. Some parishes have a rector, others a vicar.

The term has also been re-introduced to designate the priest in charge of a team ministry. (See also curate)

In the Church of Ireland, Scottish Episcopal Church and Anglican Church of Canada, most parish priests are called rectors, not vicars. However, in the some dioceses of the Anglican Church of Canada rectors are officially licensed as incumbents to express the diocesan polity of employment of clergy.

In the Episcopal Church in the United States of America, the "rector" is the priest elected to head a self-supporting parish. A priest who is appointed by the bishop to head a parish in the absence of a rector is termed a "priest-in-charge", as is a priest leading a mission (that is, a congregation which is not self-supporting). "Associate priests" are priests hired by the parish to supplement the rector in his or her duties while "assistant priests" are priests resident in the congregation who help on a volunteer basis. The positions of "vicar" and "curate" are not recognized in the canons of the entire church. However, some diocesan canons do define "vicar" as the priest-in-charge of a mission; and "curate" is often used for assistants, being entirely analogous to the English situation.[12]

In schools affiliated with the Anglican church the title "rector" is sometimes used in secondary schools and boarding schools, where the headmaster is often a priest.

Protestant churches

In many Protestant congregational churches such as Baptist, Disciples of Christ, United Church of Christ, Evangelical Free Churches, etc., the rector is the person elected to lead the congregation with pastoral duties affixed to their administrative job.

Rectorates in politics and administration

Roman

Rector provinciae was the Latin generic term for the governor of a Roman province, known after the time of Suetonius, and specifically a legal term (as used in the Codices of the Emperors Theodosius I and Justinian I) after Emperor Diocletian's Tetrarchy (when they came under the administrative authority of the Vicarius of a diocese and these under a Pretorian prefect), regardless of what their specific titles (of different rank, such as Consularis, Corrector provinciae, Praeses and Proconsul) may have been.

Adriatic

A similar gubernatorial use or as Chief magistrate in city states in the Adriatic, also in the Italian form *Rettore*, includes:

- The Republic of Ragusa (presently Dubrovnik, in Croatian Dalmatia), was governed by a *Rettore* repeatedly:
  - 1190 - 1194 between the period of sovereignty of the Norman Kingdom of "Sicily" (Naples) and Venetian sovereignty, annually elected, alongside the title 'Comes';
  - 1370 - 1808, alongside the title Duke or its Slavonic equivalent Knez, during periods of sovereignty of the Hungarian crown until 1458, then the Ottoman Sultan (formally 1526 - 1718), after 1684 under the joint 'protection' of Habsburg Austria and the Ottoman Empire, then from 1798 under Austrian (and from 1806 under French) occupation until it was incorporated into Napoleonic Illyria
  - one more Rector, from 18–29 January 1814, was Simone, conte de Giorgi, the last previous incumbent, during the short-lived restoration of the republic
- *Primo Rettore*, from 8 September 1920 to 29 December 1920, Gabriele D'Annunzio (born 1863 - died 1938) (formerly Italian Commander) in Fiume

Other

- For the use of the style *duke and rector of Burgundy* by the Zähringer dynasty claimants to viceregal powers as Regent in the Arelat kingdom of Burgundy *within* the Holy Roman Empire, see King of Burgundy#Rectorate of Burgundy
- Contemporary charters in Latin used a number of additional styles for the Danish king Cnut (Canute the Great, with Norway as his third realm; 23 April 1016 - 12 November 1035 in Britain) having *rex Anglorum* in the core plus various other titles, including *rex Anglorum totiusque Brittannice orbis gubernator et rector* i.e. 'king of the Angli and of all Britain governor and rector' (the last two in the generic sense 'ruler')
- In an early 12th-century oath to Ramon Berenguer III, Count of Barcelona, this ruler is referred to as *rector catalanicus* (as well as *catalanicus heroes* and *dux catalanensis*).
- The Comtat Venaissin in southern France was administered by a Rector since it became a papal possession until 1790 (on 24 May its States-General (representative assembly) proclaimed a constitution, but remained loyal to the pope).
- In a few 'Crown lands' of the Austrian Empire, one seat in the *Landtag* (regional legislature of semi-feudal type) was reserved for the Rector of the capital's university, notably: Graz in Steiermark (Styria), Innsbruck in Tirol, Wien (Vienna) in Nieder-Österreich (Lower Austria); in Bohemia, two Rectors seated in the equivalent Landesvertretung

Compound titles

To a rector who has resigned is often given the title *rector emeritus*. One who temporarily performs the functions usually fulfilled by a rector is styled a *pro-rector* (in parishes, administrator).

Deputies of rectors in institutions are known as *vice-rector*s (in parishes, as curates, assistant - or associate rectors, etc.). In some universities the title vice-rector has, like Vice-Chancellor in many Anglo-Saxon cases, been used for the de facto head when the essentially honorary title of rector is reserved for a high externa dignitary; until 1920, there was such a *vice-recteur* at the Parisian Sorbonne as the French Minister of Education was its nominal *Recteur*

See also

- Chancellor (education)
- Dean (education)

Notes

[1] European nations where the word "rector" is used in referring to university administrators include Albania, the Benelux, Bosnia and Herzegovina, Bulgaria, Croatia, Cyprus, Czech Republic, Denmark, Estonia, Germany, Greece, Hungary, Iceland, Italy, Latvia, Macedonia, Malta, Moldova, Poland, Portugal, Romania, Russia, Scandinavia, Scotland, Serbia, Slovenia, Slovakia, Spain, Turkey and Ukraine.

[2] "Rector" is used for university administrators in Latin American nations such as: Argentina, Bolivia, Brazil, Chile, Colombia, Paraguay, Puerto Rico, Cuba, Dominican Republic, Guatemala, Mexico, Peru and Venezuela.

[3] http://www.cam.ac.uk/univ/works/chancellor.html University of Cambridge - How the University works - The Chancellor

[4] http://www.ox.ac.uk/about_the_university/oxford_people/key_university_officers/chancellor.html Oxford University - Key University Officers - The Chancellor

[5] http://www.st-andrews.ac.uk/media/court_rector_role.pdf

[6] http://archive.gg.ca/heraldry/pub-reg/project.asp?lang=e&ProjectID=511 Badge of Office

[7] Welch, Lewis *et al.* (1899). *Yale, Her Campus, Class-rooms, and Athletics*, p. 445. (http://books.google.com/books?id=V8wWAAAAIAAJ&pg=PA301&dq=Yale+and+Noah+Porter&lr=&client=firefox-a#PPA445,M1)

[8] Ponsonby-Fane,, Richard A.B. (1956). *Kyoto: the Old Capital, 794-1869.* p. 418.

[9] Brownlee, John S. (1999). *Japanese Historians and the National Myths, 1600-1945: The Age of the Gods and Emperor Jinmu*, p. 218 n14; (http://books.google.com/books?id=eatISvupicUC&pg=PP1&dq=john+s+brownlee&client=firefox-a#PPA218,M1) N.b., Brownlee mis-identifies *Nihon Ōdai Ichiran* publication date as 1663 rather than 1657.

[10] Brownlee, John. (1991). *Political Thought in Japanese Historical Writing: From Kojiki (712) to Tokushi Yoron (1712)*, p. 120. (http://books.google.com/books?lr=&client=firefox-a&id=YSN1AAAAMAAJ&dq=john+s+brownlee&q=Honcho+tsugan&pgis=1#search)

[11] http://www.nationalarchives.gov.uk/catalogue/RdLeaflet.asp?sLeafletID=223

[12] *Canons of the Episcopal Church in the United State of America*, III.9.3

References

- 🌐 "Rector". *Catholic Encyclopedia*. New York: Robert Appleton Company. 1913.

External links

- Austria-Hungary Empire (http://www.donaumonarchie.com/) in German (use English and French translations with due caution)
- WorldStatesmen- here Croatia-Ragusa (http://www.worldstatesmen.org/Croatia.html#Ragusa) & UK (http://www.worldstatesmen.org/United_Kingdom.html)
- Chisholm, Hugh, ed. (1911). "Rector". *Encyclopædia Britannica* (11th ed.). Cambridge University Press.

Great_Brington

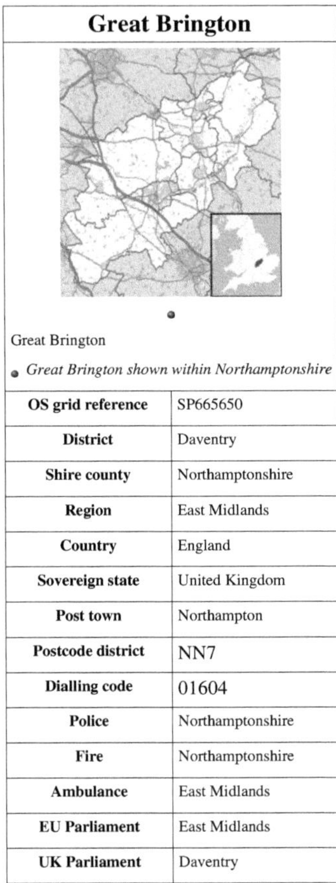

Great Brington

Great Brington

Great Brington shown within Northamptonshire

OS grid reference	SP665650
District	Daventry
Shire county	Northamptonshire
Region	East Midlands
Country	England
Sovereign state	United Kingdom
Post town	Northampton
Postcode district	NN7
Dialling code	01604
Police	Northamptonshire
Fire	Northamptonshire
Ambulance	East Midlands
EU Parliament	East Midlands
UK Parliament	Daventry

Great Brington is a village in the Daventry district of the county of Northamptonshire, England. The village, in the civil parish of Brington, has a population of about 200. The parish church is dedicated to St Mary the Virgin and St John.

Just outside the village is Althorp, the home of the Spencer family and Diana, Princess of Wales. Many conspiracy theorists believe that Diana is in fact buried at Great Brington church - as several of her relatives are, including her father the 8th Earl Spencer, who died in 1992 -[1] and not at Althorp. The death of Diana had quite an effect on the village - the village pub was renamed from *"The Fox and Hounds"* to the *"Althorp Coaching Inn"* and the previously sleepy post office gained currency exchange facilities following the large increase in tourism to the area.

The Macmillan Way long distance footpath passes through Great Brington. The disc jockey and television presenter Jo Whiley is from the village.

Geography

Nearby settlements include Little Brington, Nobottle and Long Buckby

References

[1] (http://www.bbc.co.uk/politics97/diana/althorp.html)

External links

- BBC feature on Great Brington church (http://www.bbc.co.uk/religion/religions/christianity/features/churches/grtbrington.shtml)
- Great Brington Parish Church website (http://www.greatbringtonparishchurch.co.uk)
- Map sources for Great Brington

FA_Cup

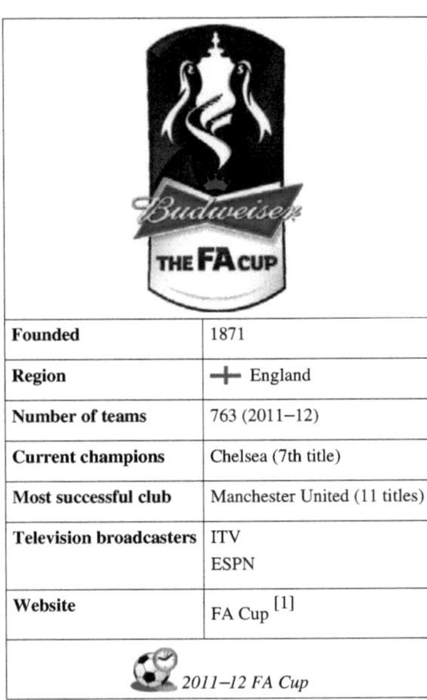

Founded	1871
Region	✚ England
Number of teams	763 (2011–12)
Current champions	Chelsea (7th title)
Most successful club	Manchester United (11 titles)
Television broadcasters	ITV ESPN
Website	FA Cup [1]

2011–12 FA Cup

The Football Association Challenge Cup, commonly known as the **FA Cup**, is a knockout cup competition in English football and is the oldest association football competition in the world.[2] The "FA Cup" is run by and named after the Football Association and usually refers to the English men's tournament, although a women's tournament is also held. Its current sponsored name is **the FA Cup with Budweiser**.[3]

The FA Cup was first held in 1871–72. Entry is open to all teams who compete in the Premier League, the Football League and in steps one to five of the FA National League System, as well as selected teams in step 6.[4] This means that clubs of all standards compete, from the largest clubs in England and Wales down to amateur village teams. The tournament has become known for the possibility for "minnows" from the lower divisions to become "giant-killers" by eliminating top clubs from the tournament and even theoretically winning the Cup, although lower division teams rarely progress beyond the early stages. The qualification rounds and a system of byes mean that the very smallest and very biggest teams almost never meet.

The holders of the FA Cup are Chelsea, who defeated Liverpool 2–1 in the 2012 final for their fourth triumph in six years and seventh overall.

Format

The competition is a knockout tournament with pairings for each round drawn at random. There are no seeds and the draw for each round is not made until after the scheduled dates for the previous round. The draw also determines which teams will play at home.

Each tie is played as a single leg. If a match is drawn, there is a replay, usually at the ground of the team who were away for the first game. Drawn replays are now settled with extra time and penalty shootouts, though until the 1990s further replays would be played until one team was victorious. Some ties took as many as six matches to settle; in their 1975 campaign, Fulham played a total of 12 games over six rounds, which remains the most games played by a team to reach a final.[5] Replays were traditionally played three or four days after the original game, but from 1991–92 they were staged at least 10 days later on police advice. This led to penalty shoot-outs being introduced. Replays are no longer held for the semi-finals or final.

There are a total of 14 rounds in the competition — six qualifying rounds, followed by six further rounds (the "proper" rounds), semi-finals, and the final. The competition begins in August with the Extra Preliminary Round, followed by the Preliminary Round and First Qualifying Round, which are contested by the lowest-ranked clubs. Clubs playing in the Conference North and Conference South are given exemption to the Second Qualifying Round, and Conference National teams are given exemption to the Fourth Qualifying Round. The 32 winners from that round join the 48 clubs from League One and League Two in the First Round (often called the First Round Proper). Finally, teams from the Premier League and Football League Championship enter at the Third Round Proper, at which point there are 64 teams remaining in the competition. The Sixth Round Proper is the quarter-final stage, at which point eight teams remain.

The qualifying rounds are regionalised to reduce the travel costs for smaller non-league sides. The First and Second Rounds were also previously split into Northern and Southern sections, but this practice was ended after the 1997–98 competition.

The FA Cup has a set pattern for when each round is played. Normally the First Round is played in mid-November, with the Second Round on one of the first two Saturdays in December. The third round is played on the first weekend in January, with the Fourth Round later in the month and Fifth Round in mid-February. The Sixth Round (or quarter-finals) traditionally occurs in early or mid March, with the semi-finals a month later. The final is normally held the Saturday after the Premier League season finishes in May. The only seasons in recent times when this pattern was not followed were 1999–2000, when most rounds were played a few weeks earlier than normal as an experiment, and 2010–2011 when the FA Cup Final was played before the Premier League season had finished, in order to allow the stadium to be ready for the UEFA Champions League final.[6]

As well as being presented with the trophy, the winning team also qualifies for the UEFA Europa League (formerly named the UEFA Cup), even if being relegated. In the event the FA Cup winners are qualified for the UEFA Champions League the FA Cup runners-up will qualify, but in the place reserved for the lowest ranking Barclays Premier League representative. The place with direct access to the Group Stage will be given to the club that finishes in the highest position in the Barclays Premier League of the clubs qualifying for the UEFA Europa League. [7] The FA Cup winners also qualify for the single-match FA Community Shield against the Premier League Champions.

The draw

The draw for each of the "proper" rounds is not seeded and is broadcast live on television, usually taking place at the conclusion of live coverage of one of the games of the previous round. No teams are seeded in the qualifying round draws either, but the teams are grouped geographically in the qualifying rounds to reduce travel costs. Public interest is particularly high during the draw for the third round, which is where the top-ranked teams are added to the draw. Traditionally, the draw has been conducted by means of balls drawn from a purple velvet bag, however in recent years in order to comply with FIFA regulations the balls have been drawn from a clear perspex container . To keep with tradition, however, the presenters are shown emptying the balls from the old velvet bag into the perspex container before the draw.

Eligible teams

All clubs in the Premier League and Football League are automatically eligible, and clubs in the next six levels of the English football league system are also eligible provided they have played in either the FA Cup, FA Trophy or FA Vase competitions in the previous season. Newly formed clubs that start playing in a high league, such as F.C. United of Manchester and MK Dons, may not therefore play in the FA Cup in their first season. All clubs entering the competition must also have a suitable stadium. There are also other factors and measures in place for lower clubs to enter the competition. It is very rare for top clubs to miss the competition, although it can happen in exceptional circumstances. Manchester United withdrew from the 1999–2000 competition due to their participation in the FIFA Club World Championship, a move highly controversial at the time.[8] [9]

Welsh sides that play in English leagues are eligible, although since the creation of the League of Wales there are only six clubs remaining: Cardiff City (the only non-English team to win the tournament, in 1927), Swansea City, Wrexham, Merthyr Town, Newport County and Colwyn Bay. In the early years other teams from Wales, Ireland and Scotland also took part in the competition, with Glasgow side Queen's Park reaching the final in 1884 and 1885 before being barred from entering by the Scottish Football Association.

The number of entrants has increased greatly in recent years. In the 2004–05 season, 660 clubs entered the competition, beating the long-standing record of 656 from the 1921–22 season. In 2005–06 this increased to 674 entrants, in 2006–07 to 687, in 2007–08 to 731 clubs, and for the 2008–09 and 2009–10 competitions it reached 762.[10] By comparison, the other major English domestic cup, the League Cup, involves only the 92 members of the Premier League and Football League.

Venues

Matches in the FA Cup are usually played at the home ground of one of the two teams. The team who plays at home is decided when the matches are drawn. There is no seeding system in place within rounds other than when teams enter the competition, therefore the home team is simply the first team drawn out for each fixture. Occasionally games may have to be moved to other grounds due to other events taking place, security reasons or a ground not being suitable to host popular teams. In the event of a draw, the replay is played at the ground of the team who originally played away from home, with a penalty shootout deciding the winner if the replay game also ends in a tie.

In the days when multiple replays were possible, the second replay (and any further replays) were played at neutral grounds. The clubs involved could alternatively agree to toss for home advantage in the second replay.

Traditionally, the FA Cup Final was played at London's Wembley Stadium. Early finals were played in other locations and, due to extensive redevelopment of Wembley, finals between 2001 and 2006 were played at Millennium Stadium in Cardiff. The final returned to Wembley in May 2007.[11] Early finals venues include Kennington Oval, in 1872 and 1874–92, the Racecourse Ground, Derby in 1886, Fallowfield Stadium, Manchester in 1893, Goodison Park in 1894, Burnden Park for the 1901 replay, Bramall Lane in 1912, the Crystal Palace Park, 1895–1914, Stamford Bridge 1920–22, and Lillie Bridge, Fulham, London in 1873. In more recent times the memorable 1970 final replay between Leeds and Chelsea was held at Old Trafford in Manchester. This was the only time between 1923 and 2000 that the FA Cup final or the FA Cup Final replay was held at a stadium other than Wembley.

The semi-finals were traditionally contested at high-capacity neutral venues; usually the home grounds of teams not involved in that semi-final. Venues used since 1990 include Manchester City's now demolished Maine Road stadium, Manchester United's Old Trafford Stadium, Sheffield Wednesday's Hillsborough stadium, Arsenal's former home, Highbury (since redeveloped as housing), London's Wembley Stadium, the Millennium Stadium in Cardiff and Aston Villa's Villa Park in Birmingham. Villa Park is the most used stadium, with 55 semi-finals. The 1991 semi-final between Arsenal and Tottenham was the first to be played at Wembley, as were both 1993, 1994 and 2000 semi-finals. In 2005, both were held at the Millennium Stadium. The decision to hold the semi-finals at the same location as the final can be controversial amongst fans[12] However, starting with the 2008 cup, all semi-finals are played at Wembley; the stadium was not ready for the 2007 semi-finals. For a list of semi-final results and the venues used, see FA Cup Semi-finals.

Trophies

The second FA Cup trophy, used between 1896 and 1910.

At the end of the final, the winning team is presented with a trophy, also known as the "FA Cup", which they hold until the following year's final. Traditionally, at Wembley finals, the presentation is made at the Royal Box, with players, led by the captain, mounting a staircase to a gangway in front of the box and returning by a second staircase on the other side of the box. At Cardiff the presentation was made on a podium on the pitch.

The cup is decorated with ribbons in the colours of the winning team; a common quiz question asks, "What is always taken to the Cup Final, but never used?" (the answer is "the losing team's ribbons"). However this is not entirely true, as during the game the cup actually has both teams' sets of ribbons attached and the runners-up ribbons are removed before the presentation. Individual members of the teams playing in the final are presented with winners' and runners'-up medals. The present FA Cup trophy is the fourth.

The first, the 'little tin idol', was used from the inception of the Cup in 1871–72 until it was stolen from a Birmingham shoe shop window belonging to William Shillcock while held by Aston Villa on 11 September 1895, and was never seen again. The FA fined Villa £25 to pay for a replacement. Almost 60 years later, the thief admitted that the cup had been melted down to make counterfeit half-crown coins.[13]

The second trophy was a replica of the first, and was last used in 1910 before being presented to the FA's long-serving president Lord Kinnaird. It was sold at Christie's on 19 May 2005 for £420,000 (£478,400 including auction fees and taxes) to David Gold, the then joint chairman of Birmingham City. David Gold has loaned this trophy to the National Football Museum which is housed in Preston North End's Deepdale Stadium and it is on permanent display to the public.

A new, larger, trophy was bought by the FA in 1911 designed and manufactured by Fattorini's of Bradford and won by Bradford City in its first outing, the only time a team from Bradford has reached the final. This trophy still exists but is now too fragile to be used, so an exact replica was made by Toye, Kenning and Spencer[14] and has been in use since the 1992 final. A "backup" trophy was made alongside the existing trophy in 1992, but it has not been used so far, and will only be used if the current trophy is lost, damaged or destroyed. An otherwise identical, but smaller replica was also made by Fattorini, the North Wales Coast FA Cup trophy, and is contested annually by members of that regional Association.

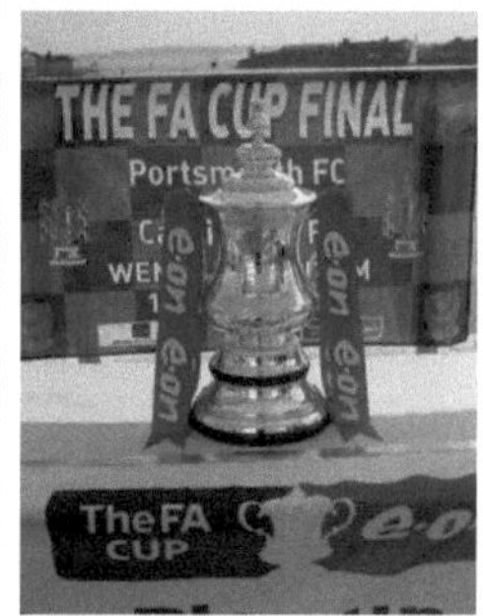

The current FA trophy, used since 1911

Though the FA Cup is the oldest domestic football competition in the world, its *trophy* is not the oldest; that title is claimed by the Youdan Cup. The oldest national trophy is the Scottish Cup.

In 1914 Burnley F.C. won the cup and received unique medals incorrectly struck as "English Cup Winners". One is displayed at Turf Moor, within the 1914 collection.

Sponsorship

Since the start of the 1994–95 season, the FA Cup has been sponsored. However, to protect the identity of the famous competition, the name has never changed from 'The FA Cup', unlike sponsorship deals for the League Cup. Sponsorship deals run for four years, though – as in the case of E.ON – one-year extensions may be agreed.

Below is a list of sponsors and the sponsored name of the competition:

Period	Sponsor	Name
1871–1994	No main sponsor	**The FA Cup**
1994–1998	Littlewoods Pools	**The FA Cup sponsored by Littlewoods**[15]
1998–2002	AXA	**The AXA-Sponsored FA Cup**[16]
2002–2006	No main sponsor	**The FA Cup**
2006–2011	E.ON	**The FA Cup sponsored by E.ON**[17] [18]
2011–2014	Budweiser	**The FA Cup with Budweiser**[19]

From August 2006 to 2014, Umbro will supply match balls for all FA Cup matches.

Giant-killers

Aside from the non-top-flight winners mentioned below, the FA Cup has a long tradition of lower-ranked teams becoming "giant-killers" by defeating opponents from a higher division.[20] While it is common for this to happen (one statistical analysis based on four years of results showed that the probability of at least one team beating one from a higher division in a given year was 99.85%, dropping to 48.8% for a two-division gap and 39.28% for a three-division gap[21]), it is considered particularly newsworthy when the "victim" is one of the top Premier League teams, or where the giant-killer is from outside the League divisions. The most recent example of a non-league team beating top-flight opposition was Sutton United's victory over Coventry City in 1988–89. The last time a non-league side beat top-flight opposition away from home was Altrincham's victory over Birmingham City at St. Andrew's in 1985–86. Within the football league one of the most notable results was Wrexham's victory over Arsenal in 1992. In the previous season, Wrexham had finished 92nd (last) in the football league, Arsenal were 1st.

Giant-killings of various scales happen every year: almost every club in the League Pyramid has a fondly remembered "giant-killing" act in its history and some small clubs have, whether by accident or design, gained a reputation for being "cup specialists" after two or more such feats within a few years.[21] Victories by non-league sides over league opposition are referred to as 'league scalps' . Overall, Yeovil Town holds the record with 20 league scalps won before the club entered the league.[22] The record for a club which has never entered the league is held by Altrincham, with 16 league scalps.

Linked to this giant-killing is the progression of teams beyond what would normally be expected. A few teams have won the FA Cup while outside of the top division, though no team from the third level of the football league has progressed to the final. For non-league teams, reaching the third round – where all top flight sides now enter – is considered a major achievement. During the 2008-09 FA Cup, a record nine teams achieved this feat,[23] and while Tottenham Hotspur won the 1901 FA Cup as a Southern League club, no non-league team has since progressed past the fifth round, this occurring most recently to Crawley Town F.C. in 2011.[24] Chasetown are the lowest ranked team to play in the third round, playing eventual runners-up Cardiff City in the 2007–08 competition. The game took place on 5 January 2008 whilst Chasetown were playing in the Southern League Division One Midlands, the eighth tier of the English football pyramid.[25]

FA Cup winners and finalists

Three clubs have won consecutive FA Cups on more than one occasion: Wanderers (1872, 1873 and 1876, 1877, 1878), Blackburn Rovers (1884, 1885, 1886 and 1890, 1891), and Tottenham Hotspur (1961, 1962 and 1981, 1982).

Seven clubs have won the FA Cup as part of a League and Cup double, namely Preston North End (1889), Aston Villa (1897), Tottenham Hotspur (1961), Arsenal (1971, 1998, 2002), Liverpool (1986), Manchester United (1994, 1996, 1999) and Chelsea (2010). Arsenal and Manchester United share the record of three doubles. Arsenal have won a double in each of three separate decades (1970s, 1990s, 2000s). Manchester United's three doubles in the 1990s highlights their dominance of English football at the time.

In 1993, Arsenal became the first side to win both the FA Cup and League Cup in the same season, beating Sheffield Wednesday 2–1, in both finals. Liverpool repeated this feat in 2001, as did Chelsea in 2007.

In 1998–99, Manchester United added the 1999 Champions League crown to their double, an accomplishment known as The European treble. Two years later, in 2000–01, Liverpool won the FA Cup, League Cup and UEFA Cup to complete a cup treble.

Portsmouth have the unusual accolade of holding the FA Cup for the longest unbroken period of time; having won the Cup in 1939, the next final was not contested until 1946, due to the outbreak of the Second World War.

The FA Cup has only been won by a non-English team once. Cardiff City achieved this in 1927 when they beat Arsenal in the final at Wembley. They had previously made it to the final only to lose to Sheffield United in 1925, and lost another final to Portsmouth in 2008.

Winners from outside the top flight

Since the foundation of the Football League, Tottenham Hotspur in 1901 have been the only non-league winners of the FA Cup. They were then playing in the Southern League and were only elected to the Football League in 1908. At that time the Football League consisted of only two 18-team divisions.

In the history of the FA Cup, only eight teams who were playing outside the top level of English football have gone on to win the competition, the most recent being West Ham United, who beat Arsenal in 1980. Excluding Tottenham in 1901, these clubs were all playing in the old Second Division, no other Third Division or lower side having reached the final.

One of the most famous upsets was when Sunderland beat Leeds United 1–0 in 1973. Leeds were third in the First Division and Sunderland were in the Second.[26] Three years later Second Division Southampton also won the Cup, against First Division Manchester United by the same 1–0 scoreline. The other non-top flight winners of the FA Cup were Notts County in 1894, the first non-top flight team to win the FA Cup since the inception of the league; Wolverhampton Wanderers in 1908; Barnsley in 1912; and West Bromwich Albion in 1931. West Bromwich Albion remain the only team to have won the FA Cup and promotion from the second flight in the same season.

Thus far the FA Cup final has never been contested by two teams from outside the top flight. Uniquely, in 2007–08, three of the four semi-finalists (Barnsley, Cardiff City and West Bromwich Albion), were from outside the top flight, although Portsmouth went on to win it.[27]

Media coverage

The FA Cup Final is one of ten events reserved for live broadcast on UK terrestrial television under the Ofcom Code on Sports and Other Listed & Designated Events.

From August 2008 until June 2012, FA Cup matches are shown live by ITV1 across England and Wales, with UTV broadcasting to Northern Ireland. ITV shows sixteen FA Cup games per season, including first pick live matches from each of the 1st to 6th rounds of the competition plus one semi-final exclusively live. The final is also shown live on ITV1.

Under the same contract, Setanta Sports showed three games and one replay in each round from round three to five, two quarter-finals, one semi-final and the final. The channel also broadcast ITV's matches exclusively to Scotland, after the ITV franchise holder in Scotland, STV, decided not to broadcast FA Cup games. Setanta entered administration in June 2009 and as a result the FA terminated Setanta's deal to broadcast the FA Cup and England internationals.[28]

In October 2009, The FA announced that ITV would show an additional match in the First and Second Rounds on ITV1, with one replay match shown on ITV4. One match and one replay match from the first two rounds will broadcast on The FA website for free, in a similar situation to the 2010 World Cup Qualifier between Ukraine and England.[29] The 2009–10 First Round match between Oldham Athletic and Leeds United was the first FA Cup match to be streamed online live.[30]

Many expected BSkyB to make a bid to show some of the remaining FA Cup games for the remainder of the 2009/10 season which would include a semi-final and shared rights to the final.[31] ESPN took over the package Setanta held for the FA Cup from the 2010/11 season.[32] The 2011 final was also shown live on Sky 3D in addition to ESPN (who provided the 3D coverage for Sky 3D) & ITV.[33]

BBC Radio Five Live provide radio coverage including several full live commentaries with additional commentaries broadcast on BBC local radio stations.

Until the 2008/09 season, the BBC and Sky Sports shared television coverage, with the BBC showing three matches in the earlier rounds. Some analysts argued the decision to move away from the Sky and, in particular, the BBC undermined the FA Cup in the eyes of the public.[34]

The FA Cup 2008–09 early rounds were being covered for the first time by ITV's online property, ITV Local. The first match of the season, between Wantage Town and Brading Town, was broadcast live online. Highlights of eight games of each round were being broadcast as catch up on ITV Local.[35] [36] Since the end of the ITV Local service, it is unknown whether or not this coverage will continue.

The FA sells overseas rights separately from the domestic contract. In Australia, FA Cup games from the 1st Round to the Semi-Finals are broadcast exclusively by Setanta Sports Australia, and the final is co-broadcast with SBS. Due to Australian anti-siphoning laws, the FA Cup Final is one of a select few sporting events of national significance, that must be shown on free-to-air commercial television. Meanwhile FOX Soccer Channel owns the rights in the United States. Supersport broadcasts the tournament in Africa, and Sony Pix in India.

See also

- FA Cup Final
- FA Cup Semi-finals
- List of FA Cup winning managers
- FA Cup records

References

[1] http://www.thefa.com/TheFACup/

[2] The oldest Cup competion [sic] in the world is at the fourth round stage, while Manchester United are in Premier League action (http://www.rte.ie/sport/soccer/2010/0122/facup.html). RTÉ. Retrieved on January 22, 2010.

[3] (http://www.thefa.com/TheFACup/FACompetitions/TheFACup/NewsAndFeatures/2011/budweiser-lead-sponsor)

[4] The FA Cup, the FA Trophy, the FA Vase and the FA Youth Cup Competitions (http://www.thefa.com/TheFACup/~/media/Files/PDF/The FA Cup And Comps/CompetitionApplications/Entry Qualifications Guidance Notes 2010-2011.ashx/Entry Qualifications Guidance Notes 2010-2011.pdf) The FA.com. Accessed 15-02-11

[5] "TheFA.com – Hammers nail Fulham" (http://www.thefa.com/TheFACup/FACompetitions/TheFACup/History/HistoryOfTheFACup/1975WestHamFulham.aspx). the FA. . Retrieved 2005-03-05.

[6] Tarnished FA Cup needs a Manchester derby's drama, The Guardian, 6 March 2011 (http://www.guardian.co.uk/football/blog/2011/mar/06/fa-cup-final-manchester-derby)

[7] (http://www.premierleague.com/en-gb/fans/faqs/who-qualifies-to-play-in-europe/) Premier League, 3 May 2012

[8] "Man Utd's FA Cup catastrophe" (http://news.bbc.co.uk/sport2/hi/football/teams/m/man_utd/853647.stm). BBC News. 27 July 2000. . Retrieved 1 March 2012.

[9] "I regret Manchester United's FA Cup pull-out: Fergie" (http://www.belfasttelegraph.co.uk/sport/football/premiership/i-regret-manchester-uniteds-fa-cup-pullout-fergie-14583761.html). *Belfast Telegraph*. 3 December 2009. . Retrieved 1 March 2012.

[10] record number of entries for 2008/9 (http://www.thefa.com/TheFACup/TheFACup/NewsAndFeatures/Postings/everyones_up_for_the_cup.htm)

[11] Nurse, Howard (2006-10-19). "Wembley Stadium to open next year" (http://news.bbc.co.uk/sport1/hi/football/6039052.stm). BBC. . Retrieved 2007-03-17.

[12] "Football supporters hail FA Cup semi final decision" (http://web.archive.org/web/20070208165141/http://www.fsf.org.uk/news/news0002-facup.html). FSF. Archived from the original (http://www.fsf.org.uk/news/news0002-facup.html) on 2007-02-08. . Retrieved 2007-02-08.

[13] The Sunday Times *Illustrated History Of Football* Reed International Books Limited. 1996. p11. ISBN 1-85613-341-9

[14] "Toye trophies page" (http://www.toye.com/products/sports/trophies-awards/). .

[15] "F.A. Cup Soccer Gets A Sponsor" (http://query.nytimes.com/gst/fullpage.html?res=9401E1DF1538F931A3575AC0A962958260). New York Times. 2 September 1994. . Retrieved 10 October 2011.

[16] "Axa wins FA Cup" (http://news.bbc.co.uk/1/hi/sport/football/138103.stm). BBC News. 23 July 1998. . Retrieved 10 October 2011.

[17] FA announces new Cup sponsorship (http://news.bbc.co.uk/sport1/hi/football/fa_cup/4676576.stm)

[18] http://www.thefa.com/TheFACup/FACompetitions/TheFACup/NewsAndFeatures/2010/eon-180610

[19] Budweiser up for Cup in £8m a year deal (http://www.thesun.co.uk/sol/homepage/sport/football/3640588/Budweiser-up-for-Cup-in-8m-a-year-deal.html)

[20] F.A. Cup Giant Killers (http://bleacherreport.com/articles/326087-fa-cup-giant-killing) Tiger, Carolina. Bleacher Report. Accessed 20–01–10

[21] https://www.timesonline.co.uk/article/0,,7973-1430225,00.html

[22] TheFA.com – Twenty to tackle answers (http://www.thefa.com/TheFACup/TheFACup/NewsAndFeatures/Postings/2005/12/FACup_TwentyToTackleAnswers.htm)

[23] Kessel, Anna (2009-01-03). "Non-league presence in third round of FA Cup breaks all-time record" (http://www.guardian.co.uk/football/2009/jan/03/fa-cup-third-round-non-league). *The Guardian* (London). . Retrieved 2010-05-02.

[24] "Ask Albert – Number 8". *BBC News*. 2000-12-07.

[25] Chasetown 1–3 Cardiff (http://news.bbc.co.uk/sport1/hi/football/fa_cup/7163953.stm).

[26] "TheFA.com – Shocks do happen" (http://web.archive.org/web/20050305044037/http://www.thefa.com/TheFACup/TheFACup/History/Postings/2003/11/46982.htm). The FA. Archived from the original (http://www.thefa.com/TheFACup/TheFACup/History/Postings/2003/11/46982.htm) on 2005-03-05. . Retrieved 2005-04-06.

[27] "FA Cup semi-final draw 2008" (http://news.bbc.co.uk/sport1/hi/football/fa_cup/7286364.stm). *BBC Sport*. British Broadcasting Corporation. 10 March 2008. . Retrieved 15 March 2012.

[28] "FA face Setanta shortfall" (http://news.bbc.co.uk/sport1/hi/football/8115805.stm). *BBC News*. 2009-06-23. . Retrieved 2009-08-12.

[29] "FA Cup to be broadcast Free-to-Air" (http://www.thefa.com/TheFACup/FACompetitions/TheFACup/NewsAndFeatures/2009/FACup_TVTies_1P.aspx). . Retrieved 2009-10-27.

[30] "Latics to face Leeds in Cup" (http://www.oldham-chronicle.co.uk/news-features/10/oldham-athletic-news/31689/latics-to-face-leeds-in-cup). . Retrieved 2009-10-27.

[31] "FA Cup and England TV rights up for grabs as Setanta falls into administration and prepares to disappear from our screens" (http://www.dailymail.co.uk/sport/football/article-1194991/FA-Cup-England-TV-rights-grabs-Setanta-falls-administration-prepares-disappear-screens.html). *Daily Mail* (London). 23 June 2009. . Retrieved 2009-12-30.

[32] Gibson, Owen (7 December 2009). "ESPN secures rights to show FA Cup matches from next season" (http://www.guardian.co.uk/football/2009/dec/07/fa-cup-espn-bbc-itv). *The Guardian* (London). . Retrieved 2009-12-30.

[33] "ESPN's 3D coverage of 2011 FA Cup Final to be available on Sky 3D" (http://corporate.sky.com/media/press_releases/2011/ESPN_3D_Coverage.htm). *Sky TV* (London). . Retrieved 2011-04-26.

[34] Sale, Charles (2009-09-02). "EXCLUSIVE: E.ON opt against extending FA Cup sponsorship deal" (http://www.dailymail.co.uk/sport/football/article-1210599/EXCLUSIVE-E-ON-opt-extending-FA-Cup-sponsorship-deal.html?ITO=1490). *Daily Mail* (London). .

[35] "Watch The FA Cup online" (http://www.thefa.com/TheFACup/TheFACup/NewsAndFeatures/Postings/FA_Cup_online.htm). .

[36] "Cup tie live online" (http://www.thefa.com/TheFACup/TheFACup/NewsAndFeatures/Postings/2008/07/Watch_live_online.htm). .

External links

- The FA Cup Archive (http://www.thefa.com/TheFACup/FACompetitions/TheFACup/Archive.aspx) − England's official Football Association site, all results with dates, including all qualifying rounds
- The official FA Cup website (http://www.thefa.com/TheFACup/)
- Thomas Fattorini Ltd. makers of the 1911 FA Cup (http://www.fattorini.co.uk) − manufacturers of the 1911 FA Cup and other sporting trophies
- FA Cup going under the hammer (http://news.bbc.co.uk/sport1/hi/football/fa_cup/4151177.stm) − BBC News story on the sale of the second trophy
- FA Supporters (http://faclub16.com/Season2011.aspx) − Independent FA Cup Supporters Club

Northamptonshire

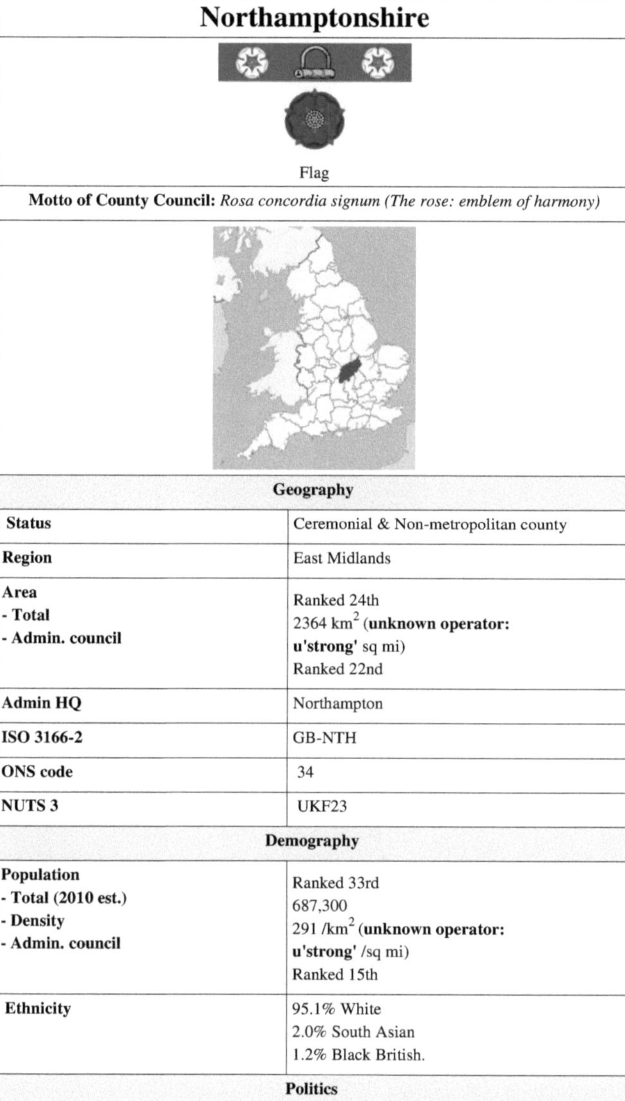

<table>
<tr><td colspan="2" align="center">Northamptonshire</td></tr>
<tr><td colspan="2" align="center">Flag</td></tr>
<tr><td colspan="2">Motto of County Council: Rosa concordia signum (The rose: emblem of harmony)</td></tr>
<tr><td colspan="2" align="center">Geography</td></tr>
<tr><td>Status</td><td>Ceremonial & Non-metropolitan county</td></tr>
<tr><td>Region</td><td>East Midlands</td></tr>
<tr><td>Area
- Total
- Admin. council</td><td>Ranked 24th
2364 km^2 (unknown operator:
u'strong' sq mi)
Ranked 22nd</td></tr>
<tr><td>Admin HQ</td><td>Northampton</td></tr>
<tr><td>ISO 3166-2</td><td>GB-NTH</td></tr>
<tr><td>ONS code</td><td>34</td></tr>
<tr><td>NUTS 3</td><td>UKF23</td></tr>
<tr><td colspan="2" align="center">Demography</td></tr>
<tr><td>Population
- Total (2010 est.)
- Density
- Admin. council</td><td>Ranked 33rd
687,300
291 /km^2 (unknown operator:
u'strong' /sq mi)
Ranked 15th</td></tr>
<tr><td>Ethnicity</td><td>95.1% White
2.0% South Asian
1.2% Black British.</td></tr>
<tr><td colspan="2" align="center">Politics</td></tr>
</table>

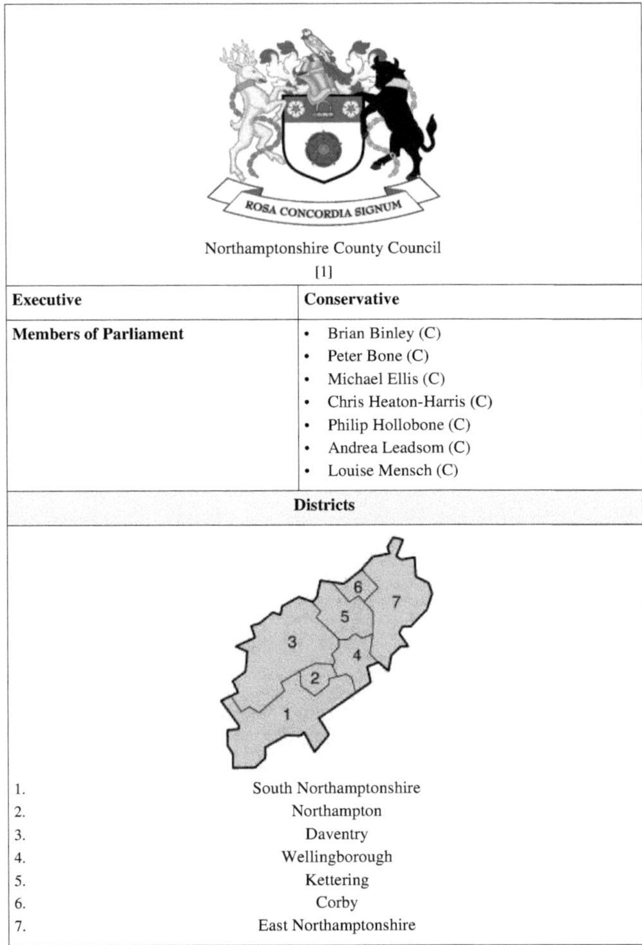

Northamptonshire County Council	
[1]	
Executive	**Conservative**
Members of Parliament	• Brian Binley (C) • Peter Bone (C) • Michael Ellis (C) • Chris Heaton-Harris (C) • Philip Hollobone (C) • Andrea Leadsom (C) • Louise Mensch (C)
Districts	

1.	South Northamptonshire
2.	Northampton
3.	Daventry
4.	Wellingborough
5.	Kettering
6.	Corby
7.	East Northamptonshire

Northamptonshire (◄ᴊ /nɔrˈθæmptənʃər/ or /nɔrθˈhæmptənʃɪər/; archaically, the **County of Northampton**; abbreviated **Northants.** or **N/hants**) is a landlocked ceremonial county in the East Midlands region of England. Its population is 629,676 as at the 2001 census. It has boundaries with eight other ceremonial counties: Warwickshire to the west, Leicestershire and Rutland to the north, Cambridgeshire to the east, Bedfordshire to the south-east, Buckinghamshire to the south, Oxfordshire to the south-west and Lincolnshire to the north-east – England's shortest county boundary at 19 metres (**unknown operator: u'strong'** yd).[2] The county seat is Northampton. Other large population centres include Kettering, Corby, Wellingborough, Rushden and Daventry.

Northamptonshire's county flower is the cowslip.

History

Much of Northamptonshire's countryside appears to have remained somewhat intractable with regards to early human occupation, resulting in an apparently sparse population and relatively few finds from the Palaeolithic, Mesolithic and Neolithic periods.[3] In about 500 BC the Iron Age was introduced into the area by a continental people in the form of the Hallstatt culture,[4] and over the next century a series of hill-forts were constructed at Arbury Camp, Rainsborough camp, Borough Hill, Castle Dykes, Guilsborough, Irthlingborough, and most notably of all, Hunsbury Hill. There are two more possible hill-forts at Arbury Hill (Badby) and Thenford.[4]

In the 1st century BC, most of what later became Northamptonshire became part of the territory of the Catuvellauni, a Belgic tribe, the Northamptonshire area forming their most northerly possession.[4] The Catuvellauni were in turn conquered by the Romans in 43 AD.[5]

The Roman road of Watling Street passed through the county, and an important Roman settlement, *Lactodorum*, stood on the site of modern-day Towcester. There were other Roman settlements at Northampton, Kettering and along the Nene Valley near Raunds. A large fort was built at Longthorpe.[4]

After the Romans left, the area eventually became part of the Anglo-Saxon kingdom of Mercia, and Northampton functioned as an administrative centre. The Mercians converted to Christianity in 654 AD with the death of the pagan king Penda.[6] From about 889 the area was conquered by the Danes (as at one point almost all of England was, except for Athelney marsh in Somerset) and became part of the Danelaw - with Watling Street serving as the boundary - until being recaptured by the English under the Wessex king Edward the Elder, son of Alfred the Great, in 917. Northamptonshire was conquered again in 940, this time by the Vikings of York, who devastated the area, only for the county to be retaken by the English in 942.[7] Consequently, it is one of the few counties in England to have both Saxon and Danish town-names and settlements.

The county was first recorded in the Anglo-Saxon Chronicle (1011), as *Hamtunscire*: the *scire* (shire) of *Hamtun* (the homestead). The "North" was added to distinguish Northampton from the other important *Hamtun* further south: Southampton - though the origins of the two names are in fact different.[8]

Rockingham Castle was built for William the Conqueror[9] and was used as a Royal fortress until Elizabethan times. The now-ruined Fotheringhay Castle was used to imprison Mary, Queen of Scots, before her execution.[10] In 1460, during the Wars of the Roses, the Battle of Northampton took place and King Henry VI was captured.[11]

George Washington, the first President of the United States of America, was born into the Washington family who had migrated to America from Northamptonshire in 1656. George Washington's great-great-great-great-great grandfather, Lawrence Washington, was Mayor of Northampton on several occasions and it was he who bought Sulgrave Manor from Henry VIII in 1539. It was George Washington's great-grandfather, John Washington, who emigrated in 1656 from Northants to Virginia. Before Washington's ancestors moved to Sulgrave, they lived in Warton, Lancashire.[12]

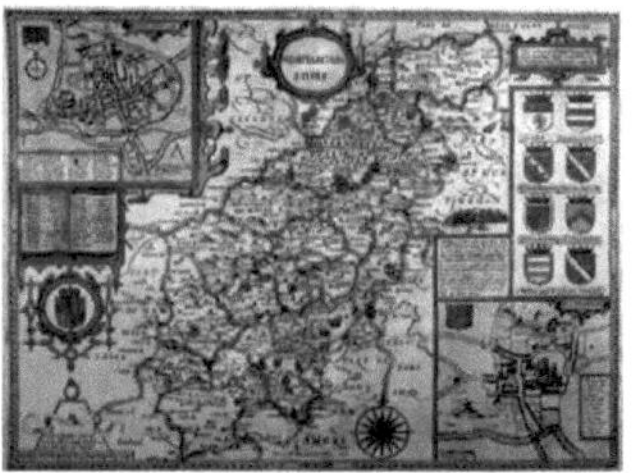

John Speed's 17th century map of Northamptonshire

During the English Civil War, Northamptonshire strongly supported the Parliamentarian cause, and the Royalist forces suffered a crushing defeat at the Battle of Naseby in 1645 in the north of the county. King Charles I was imprisoned at Holdenby House in 1647.[13]

In 1823 Northamptonshire was said to "[enjoy] a very pure and wholesome air" because of its dryness and distance from the sea. Its livestock were celebrated: "Horned cattle, and other animals, are fed to extraordinary sizes: and many horses of the large black breed are reared."[14]

Nine years later, the county was described as "a county enjoying the reputation of being one of the healthiest and pleasantest parts of England" although the towns were "of small importance" with the exceptions of Peterborough and Northampton. In summer, the county hosted "a great number of wealthy families... country seats and villas are to be seen at every step."[15] Northamptonshire is still referred to as the county of "spires and squires" because of the numbers of stately homes and ancient churches.[16]

In the 18th and 19th centuries, parts of Northamptonshire and the surrounding area became industrialised. The local specialisation was shoemaking and the leather industry and by the end of the 19th century it was almost definitively the boot and shoe making capital of the world. In the north of the county a large ironstone quarrying industry developed from 1850.[17] During the 1930s, the town of Corby was established as a major centre of the steel industry. Much of Northamptonshire nevertheless remains largely rural.

Corby was designated a new town in 1950[18] and Northampton followed in 1968.[19] As of 2005 the government is encouraging development in the South Midlands area, including Northamptonshire.[20]

Peterborough

The Soke of Peterborough was historically associated with and considered part of Northamptonshire, as the county diocese is focused upon the cathedral there.[21] However, Peterborough had its own county council, and in 1965 was merged with the neighbouring small county of Huntingdonshire.[22] Under the Local Government Act 1972 the city of Peterborough became a district of Cambridgeshire.[23]

Geography

Northampton

Kettering

Wellingborough

Corby

Daventry

Rushden

Thrapston

Brackley

Oundle

Desborough

Towcester

Irthlingborough

Kings Sutton

Brixworth

Raunds

Silverstone

Banbury

Market Harborough

Milton Keynes

Rugby

Notable places in and around Northamptonshire

Northamptonshire is a landlocked county located in the southern part of the East Midlands region[24] which is sometimes known as the South Midlands. The county contains the watershed between the River Severn and The Wash while several important rivers have their sources in the north-west of the county, including the River Nene, which flows north-eastwards to The Wash, and the "Warwickshire Avon", which flows south-west to the Severn. In 1830 it was boasted that "not a single brook, however insignificant, flows into it from any other district".[25] The highest point in the county is Arbury Hill at 225 metres (**unknown operator: u'strong'** ft).[26]

Kilworth Wharf on the Grand Union Canal

There are several towns in the county with Northampton being the largest and most populous. At the time of the 2008 estimates, a population of 685,000 lived in the county with 205,200 living in Northampton. The table below shows all towns with over 9,000 inhabitants.

Rank	Town	Population	Borough/District council
1	Northampton	205,200 (2008)	Northampton Borough Council
2	Kettering	51,063 (2001)[27]	Kettering Borough Council
3	Corby	49,222 (2001)[27]	Corby Borough Council
4	Wellingborough	46,959 (2001)[27]	Borough Council of Wellingborough
5	Rushden	25,849 (2001)	East Northamptonshire District Council
6	Daventry	22,367 (2001)	Daventry District Council
7	Brackley	13,331 (2001)	South Northamptonshire District Council

As of 2010 there are 16 settlements in Northamptonshire with a town charter:

- Brackley, Burton Latimer, Corby, Daventry, Desborough, Higham Ferrers, Irthlingborough, Kettering, Northampton, Oundle, Raunds, Rothwell, Rushden, Towcester, Thrapston and Wellingborough.

Climate

Like the rest of the British Isles, Northamptonshire has an oceanic climate (Köppen climate classification). The table below shows the average weather for Northamptonshire from the Moulton weather station.

Governance

Northamptonshire, like most English counties, is divided into a number of local authorities. The seven borough/district councils cover 15 towns and hundreds of villages. The county has a two-tier structure of local government and an elected county council based in Northampton, and is also divided into seven districts each with their own district or borough councils:[28]

Council	Where based
Corby Borough Council	Corby
Daventry District Council	Daventry
East Northamptonshire District Council	Thrapston
Kettering Borough Council	Kettering
Northampton Borough Council	Northampton
South Northamptonshire District Council	Towcester
Borough Council of Wellingborough	Wellingborough

Northampton itself is the most populous urban district in England not to be administered as a unitary authority (even though several smaller districts are unitary). During the 1990s local government reform, Northampton Borough Council petitioned strongly for unitary status, which led to fractured relations with the County Council.

Before 1974, the Soke of Peterborough was considered geographically part of Northamptonshire, although it had had a separate county council since the late 19th Century and separate Quarter Sessions courts before then. Now part of Cambridgeshire, the city of Peterborough became a unitary authority in 1998, but it continues to form part of that county for ceremonial purposes.[29]

National representation

Northamptonshire returns seven members of Parliament, who all are part of the Conservative Party.[30]

Constituency	Member of Parliament	Political party
Corby	Louise Mensch	Conservative
Daventry	Chris Heaton-Harris	Conservative
Kettering	Philip Hollobone	Conservative
Northampton North	Michael Ellis	Conservative
Northampton South	Brian Binley	Conservative
Northamptonshire South	Andrea Leadsom	Conservative
Wellingborough & Rushden	Peter Bone	Conservative

From 1993 until 2005, Northamptonshire County Council,[31] for which each of the 73 electoral divisions in the county elect a single councillor, had been held by the Labour Party; previously it had been under no overall control since 1981. The councils of the rural districts – Daventry, East Northamptonshire, and South Northamptonshire – are strongly Conservative, whereas the political composition of the urban districts is more mixed. At the 2003 local elections, Labour lost control of Kettering, Northampton, and Wellingborough, retaining only Corby. Elections for the entire County Council are held every four years – the last were held on 5 May 2005 when control of the County Council changed from the Labour Party to the Conservatives. The County Council uses a leader and cabinet executive system and abolished its area committees in April 2006.

Economy

Historically, Northamptonshire's main industry was the manufacture of boots and shoes.[32] Many of the manufacturers closed down in the Thatcher era which in turn left many county people unemployed. Although R Griggs and Co Ltd, the manufacturer of Dr. Martens, still has its UK base in Wollaston near Wellingborough,[33] the shoe industry in the county is now nearly gone. Large employers include the breakfast cereal manufacturers Weetabix, in Burton Latimer, the Carlsberg brewery in Northampton, Avon Products, Siemens, Barclaycard, Saxby Bros Ltd and Golden Wonder.[34] [35] In the west of the county is the Daventry International Railfreight Terminal;[36] which is a major rail freight terminal located on the West

Silverstone adds millions every year to the local economy - Kimi Räikkönen testing for McLaren at Silverstone in April 2006

Coast Main Line near Rugby. Wellingborough also has a smaller railfreight depot[37] on Finedon Road, called Nelisons sidings.[38]

This is a chart of trend of the regional gross value added of Northamptonshire at current basic prices in millions of British Pounds Sterling (correct on 21 December 2005):[39]

Year	Regional Gross Value Added[40]	Agriculture[41]	Industry[42]	Services[43]
1995	**6,139**	112	2,157	3,870
2000	**9,743**	79	3,035	6,630
2003	**10,901**	90	3,260	7,551

The region of Northamptonshire, Oxfordshire and the South Midlands has been described as "Motorsport Valley... a global hub" for the motor sport industry.[44] [45] The Mercedes GP[46] and Force India[47] Formula One teams have their bases at Brackley and Silverstone respectively, while Cosworth[48] and Mercedes-Benz High Performance Engines[49] are also in the county at Northampton and Brixworth.

International motor racing takes place at Silverstone Circuit[50] and Rockingham Motor Speedway;[51] Santa Pod Raceway is just over the border in Bedfordshire but has a Northants postcode.[52] A study commissioned by Northamptonshire Enterprise Ltd (NEL) reported that Northamptonshire's motorsport sites attract more than 2.1 million visitors per year who spend a total of more than £131 million within the county.[53]

Milton Keynes and South Midlands Growth area

Northamptonshire forms part of the Milton Keynes and South Midlands Growth area which also includes Milton Keynes, Aylesbury Vale and Bedfordshire. This area has been identified as an area which is due to have tens of thousands additional homes built between 2010-2020. In North Northamptonshire (Boroughs of Corby, Kettering, Wellingborough and East Northants), over 52,000 homes are planned or newly-built and 47,000 new jobs are also planned.[54] In West Northamptonshire (boroughs of Northampton, Daventry and South Northants), over 48,000 homes are planned or newly-built and 37,000 new jobs are planned.[55] To overlook the planned developments, two urban regeneration companies have been created: North Northants Development Company (NNDC)[54] and the West Northamptonshire Development Corporation.[55] The NNDC launched a controversial[56] campaign called *North Londonshire* to attract people from London to the county.[57] There is also a county-wide tourism campaign with the slogan *Northamptonshire, Let yourself grow.*[58]

Education

Northamptonshire County Council operates a complete comprehensive system with 42 state secondary schools.[59] The county's music and performing arts service provides peripatetic music teaching to schools. It also supports 15 local Saturday morning music and performing arts centres around the county[60] and provides a range of county-level music groups.[61]

Colleges

There are seven colleges across the county, with the Tresham College of Further and Higher Education having four campuses in three towns: Corby, Kettering and Wellingborough.[62] Tresham provides further education and offers vocational courses, GCSEs and A Levels.[63] It also offers Higher Education options in conjunction with several universities.[64] Other colleges in the county are: Fletton House, Knuston Hall, Moulton College, Northampton College, Northampton New College and The East Northamptonshire College.

University

Northamptonshire has one University, the University of Northampton. It has two campuses 2.5 miles (**unknown operator: u'strong'** km) apart and 10,000 students.[65] It offers courses for needs and interests from foundation and undergraduate level to postgraduate, professional and doctoral qualifications. Subjects include traditional arts, humanities and sciences subjects, as well as entrepreneurship, product design and advertising.[66]

Healthcare

Hospitals

Northampton has several National Health Service branches, the main acute NHS hospitals in the county being Northampton and Kettering General Hospitals. In the south-west of the county, the town of Brackley and surrounding villages are serviced by the Horton General Hospital in Banbury in neighbouring Oxfordshire for acute medical needs. A similar arrangement is in place for the town of Oundle and nearby villages, served by Peterborough District Hospital.

In February 2011 a new satellite out-patient centre opened at Nene Park, Irthlingborough to provide over 40,000 appointments a year, as well as a minor injury unit to serve Eastern Northamptonshire. This was opened to relieve pressure off Kettering General Hospital, and has also replaced the dated Rushden Memorial Clinic which provided at the time about 8,000 appointments a year, when open. [67]

Water contamination

In June 2008, Anglian Water found traces of Cryptosporidium in water supplies of Northamptonshire. The local reservoir at Pitsford was investigated and a European Rabbit which had strayed into it was found,[68] causing the problem. About 250,000 residents were affected;[69] by 14 July 2008, 13 cases of cryptosporidiosis attributed to water in Northampton had been reported.[70] Following the end of the investigation, Anglian Water lifted its boil notice for all affected areas on 4 July 2008.[71] Anglian Water revealed that it will pay up to £30 per household as compensation for customers hit by the water crisis.[72]

Transport

The gap in the hills at Watford Gap meant that many south-east to north-west routes passed through Northamptonshire. The Roman Road Watling Street (now part of the A5) passes through here, as did later canals, railways and major roads.

Roads

Major national roads including the M1 motorway (London to Leeds) and the A14 (Rugby to Ipswich), provide Northamptonshire with transport links, both north–south and east–west. The A43 joins the M1

Brackley bypass on the A43

to the M40 motorway, passing through the south of the county to the junction west of Brackley, and the A45 links Northampton with Wellingborough and Peterborough.

The county road network, managed by Northamptonshire County Council includes the A45 west of the M1 motorway, the A43 between Northampton and the county boundary near Stamford, the A361 between Kilsby and Banbury (Oxon) and all B, C and Unclassified Roads. Since 2009 these highways have been managed on behalf of the county council by MGWSP, a joint venture between May Gurney and WSP.

Rivers and canals

Further information: Category:Rivers of Northamptonshire

Two major canals – the Oxford and the Grand Union – join in the county at Braunston. Notable features include a flight of 17 locks on the Grand Union at Rothersthorpe, the canal museum at Stoke Bruerne, and a tunnel at Blisworth which, at 2813 metres (**unknown operator: u'strong'** yd), is the third-longest navigable canal tunnel on the UK canal network.

The Grand Union Canal at Braunston

A branch of the Grand Union Canal connects to the River Nene in Northampton and has been upgraded to a "wide canal" in places and is known as the *Nene Navigation*. It is famous for its guillotine locks.

Railways

Two trunk railway routes, the Midland Main Line and the West Coast Main Line, cross the county. At its peak, Northamptonshire had 75 railway stations. It now has only six, at Northampton and Long Buckby on the West Coast Main Line, Kettering, Wellingborough and Corby on the Midland Main Line, along with King's Sutton, which is a few metres from the boundary with Oxfordshire on the Chiltern Main Line.

A East Midlands Trains service approaching Wellingborough on the Midland Main Line

Before nationalisation of the railways in 1948 and the creation of British Railways, three of the "Big Four" railway companies operated in Northamptonshire: the London, Midland and Scottish Railway, London and North Eastern Railway and Great Western Railway. Only the Southern Railway was not represented. As of 2010 it is served by Virgin Trains, London Midland, Chiltern Railways and East Midlands Trains.

Corby rail history

Corby was described as the largest town in Britain without a railway station.[73] The railway running through the town from Kettering to Oakham in Rutland was previously used only by freight traffic and occasional diverted passenger trains that did not stop at the station. The line through Corby was once part of a main line to Nottingham through Melton Mowbray, but the stretch between Melton and Nottingham was closed in 1968. In the 1980s, an experimental passenger shuttle service ran between Corby and Kettering but was withdrawn a few years later.[74] On 23 February 2009, a new railway station opened, providing direct hourly access to London St Pancras. Following the opening of Corby Station, Rushden then became the largest town in the UK without a direct railway station.

Closed lines and stations

Railway services in Northamptonshire were reduced by the Beeching Axe in the 1960s.[75] Closure of the line connecting Northampton to Peterborough by way of Wellingborough, Thrapston, and Oundle left eastern Northamptonshire devoid of railways. Part of this route was reopened in 1977 as the Nene Valley Railway. A section of one of the closed lines, the Northampton to Market Harborough line, is now the Northampton & Lamport heritage railway, while the route as a whole forms a part of the National Cycle Network, as the Brampton Valley Way.

As early as 1897 Northamptonshire would have had its own Channel Tunnel rail link with the creation of the Great Central Railway, which was intended to connect to a tunnel under the English Channel. Although the complete project never came to fruition, the rail link through Northamptonshire was constructed, and had stations at Charwelton, Woodford Halse, Helmdon and Brackley. It became part of the London and North Eastern Railway in 1923 (and of British Railways in 1948) before its closure in 1966.

Future

In June 2009 the Association of Train Operating Companies (ATOC) recommended opening a new station on the former Irchester railway station site for Rushden, Higham Ferrers and Irchester, called Rushden Parkway.[76] Network Rail is looking at electrifying the Midland Main Line north of Bedford.[77] A open access company has approached Network Rail for services to Oakham in Rutland to London via the county.[77]

The Rushden, Higham and Wellingborough Railway would like to see the railway fully reopen between Wellingborough and Higham Ferrers. As part of the government-proposed High Speed 2 railway line (between London and Birmingham), the High speed railway line will go through the southern part of the county but with no station built.

Buses

Most buses are operated by Stagecoach in Northants and First Northampton. Some town area routes have been named the Corby Star, Connect Kettering, Connect Wellingborough and Daventry Dart; the last three of these routes have route designations that include a letter, such as A, D1, W1, W2, and so on.[78] [79]

Airports

Sywell Aerodrome, on the edge of Sywell village, has three grass runways and one concrete all weather runway. It is however only 1000 metres and therefore cannot be served by passenger jets as of yet.[80]

Sywell Aerodrome

Media

Newspapers

The two main newspapers in the county are the Northamptonshire Evening Telegraph and the Northampton Chronicle & Echo.

Television

BBC regions

Most of Northamptonshire is served by the BBC's East region which is based in Norwich. The regional news television programme, **BBC Look East**, provides local news across the East of England, Milton Keynes and most of Northamptonshire. An opt-out in *Look East* covers

BBC Radio Northampton's Broadcasting House

the west part of the region only, broadcast from Cambridge. This area also is covered by the BBC's **The Politics Show: East** and **Inside Out: East**. A small part of the northern part of the county is covered by BBC East Midlands's regional news **BBC East Midlands Today**, while a small part of South Northamptonshire is covered by BBC Oxford's regional news **BBC Oxford News** which is part of the BBC South Today programme.

ITV regions

Most of Northamptonshire is covered by ITV's Anglia region (which broadcasts **Anglia Today/Tonight**); in the south-west of the county, primarily Brackley and the surrounding villages, broadcasts can be received from the Oxford transmitter which broadcasts ITV Meridian's **Meridian Today/Tonight**.

Radio

BBC Radio Northampton, broadcasts on two FM frequencies: 104.2 MHz for the south and west of the county (including Northampton and surrounding area) and 103.6 MHz for the north of the county (including Kettering, Wellingborough and Corby). BBC Radio Northampton is located in Abington Street, Northampton. These services are broadcast from the Sandy Heath transmitter in Bedfordshire.

There are three commercial radio stations in the county. The former *Kettering and Corby Broadcasting Company (KCBC)* station is now called Connect Radio (97.2 and 107.4 MHZ FM), following a merger with the Wellingborough-based station of the same name. While both Heart Northants (96.6 MHz FM) and AM station Gold (1557 kHz) air very little local content as they form part of a national network. National digital radio is also available in Northamptonshire, though coverage is limited.

Sport

Rugby Union

Northamptonshire has many rugby union clubs. Its premier team, Northampton Saints, competes in the Aviva Premiership and won the European championship in 2000 by defeating Munster for the Heineken Cup, 9-8. Saints are based at the 13,600 capacity Franklin's Gardens ground.

Statue inscribed 'They tackled the job' outside Franklin's Gardens

Football

Northamptonshire has several football teams, the most prominent being the League Two side Northampton Town. Other football teams include Kettering Town who are in the Conference National & Corby Town, who are in the Conference North. Wellingborough Town claims to be the sixth oldest club in the country.

Cricket

Northamptonshire County Cricket Club is in Division Two of the County Championship. Northamptonshire Cricket Club has recently signed overseas professionals such as Sourav Ganguly.

Motor Sport

Silverstone is a major motor racing circuit, most notably used for the British Grand Prix. There is also a dedicated radio station for the circuit which broadcasts on 87.7 FM or 1602 MW when events are taking place. Rockingham Speedway Corby is the largest stadium in the UK with 130,000 seats. It is a US-style elliptical racing circuit (the largest of its kind outside of the US), and is used extensively for all kinds of motor racing events. The Santa Pod drag racing circuit, venue for the FIA European Drag Racing Championships is just across the border in Bedfordshire but has a NN postcode.

Swimming

There are five main swimming clubs in the county: Wellingborough, Northampton, Kettering, Daventry and Rushden. They participate in many competitions. There is also an Olympic sized swimming pool at Corby opened in 2010.

Culture

Rock and pop bands originating in the area have included Bauhaus, The Departure, New Cassettes, Raging Speedhorn and Defenestration.

Places of interest

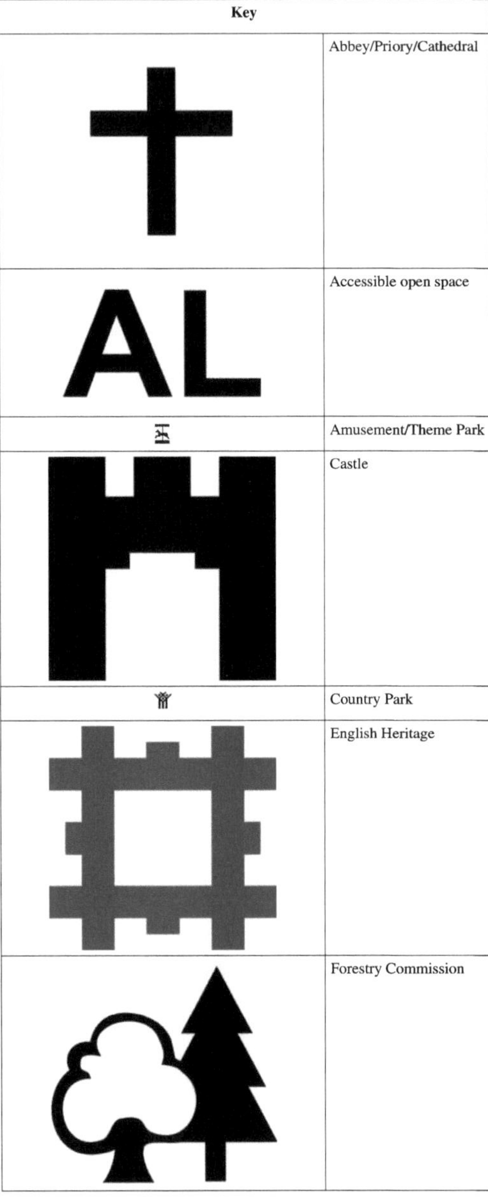

Key	
	Abbey/Priory/Cathedral
	Accessible open space
	Amusement/Theme Park
	Castle
	Country Park
	English Heritage
	Forestry Commission

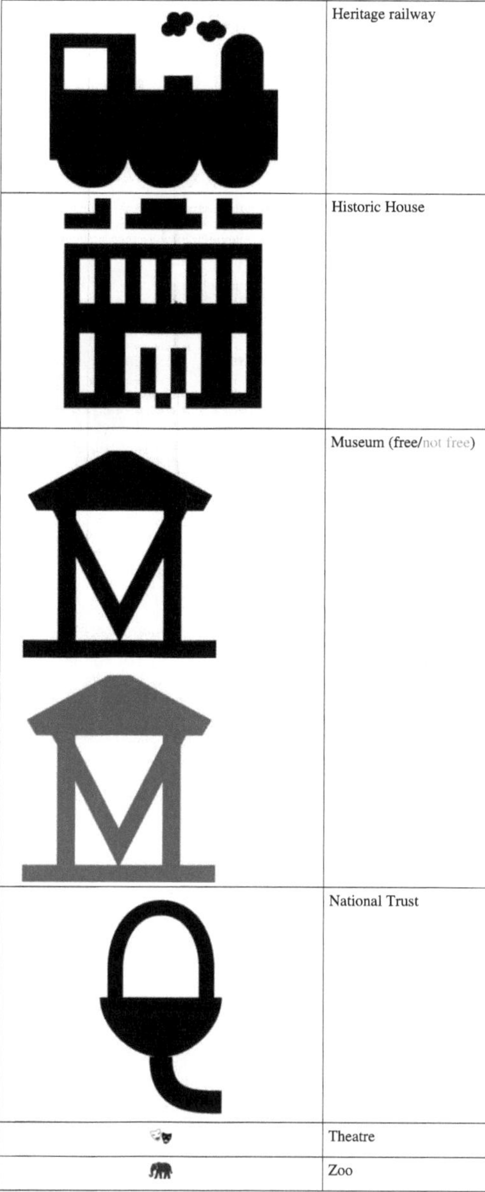

	Heritage railway
	Historic House
	Museum (free/not free)
	National Trust
	Theatre
	Zoo

- 78 Derngate
- Althorp
- Barnwell Country Park
- Barnwell Manor
- Billing Aquadrome
- Borough Hill Daventry (Iron Age hill fort) **AL**
- Boughton House (home of the Dukes of Buccleuch)
- Blisworth tunnel
- Brackley

- Knuston Hall
- Lamport Hall
- lilford Hall
- Lyveden New Bield
- Pitsford Reservoir
- Prebendal Manor House, Nassington
- Naseby Field
- Northampton Cathedral
- Northampton & Lamport Railway

- Brampton Valley Way (linear park on a disused railway line) **AL**
- Northamptonshire Ironstone Railway

- Brixworth Country Park
- Burghley House (in the Soke of Peterborough, so formerly in Northants),
- Canons Ashby House
- Castle Ashby (home of the Marquess of Northampton),
- Coton Manor Garden
- Cottesbrooke Hall
- Daventry Country Park

- Roadmender - live music venue [81]
- Piddington Roman Villa
- Rockingham Castle
- Rockingham Forest
- Rockingham Motor Speedway
- Rushden Hall
- Rushden, Higham and Wellingborough Railway

- Deene Park
- Rushden Station Railway Museum

- Delapré Abbey

- Derngate and Royal Theatre
- Easton Neston 🏛
- Elton Hall 🏛
- Fermyn Woods Country Park
- Fotheringhay Castle & Church
- Franklin's Gardens
- Geddington's Eleanor cross
- Holdenby House 🏛
- Irchester Country Park
- Jurassic Way (long-distance footpath)
- Kelmarsh Hall 🏛

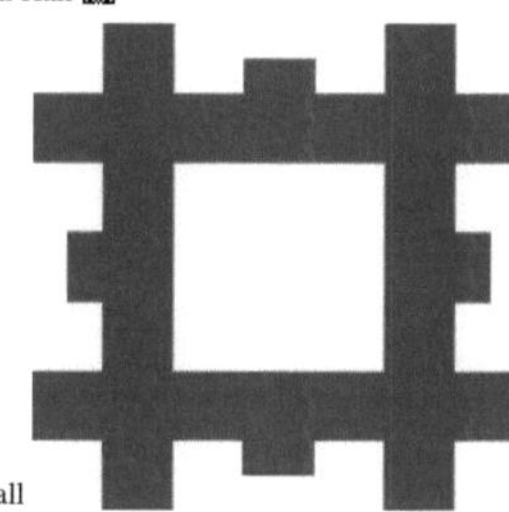

- Kirby Hall

- Rushton Triangular Lodge
- Salcey Forest
- Silverstone Circuit
- Southwick Hall 🏛
- Stanwick Lakes
- Stoke Bruerne Canal Museum
- Sulgrave Manor 🏛
- Summer Leys nature reserve
- Syresham
- Sywell Country Park
- The Castle Theatre
- Watford Locks
- Wellingborough Museum

- Whittlewood Forest
- Wicksteed Park

Annual events

- Gretton Barn dance
- British Grand Prix at Silverstone
- Burghley Horse Trials
- Crick Boat Show
- Hollowell Steam Rally
- Northampton Balloon Festival
- Rothwell Fair
- Rushden Cavalcade
- St Crispin Street Fair
- Wellingborough Carnival
- World Conker Championships

See also

- List of Lord Lieutenants of Northamptonshire
- List of High Sheriffs of Northamptonshire
- Custos Rotulorum of Northamptonshire - List of Keepers of the Rolls
- Northamptonshire (UK Parliament constituency) - Historica list of MPs for the Northamptonshire constituency
- List of places in Northamptonshire
- History of Northamptonshire
- East Midlands
- South Midlands
- Category:People from Northamptonshire

Notes

[1] http://www.northamptonshire.gov.uk/

[2] "Lincolnshire County Council" (http://www.thebythams.org.uk/localgovernment/lincolnshire-cc/index.html). Thebythams.org.uk. 2005-10-24. . Retrieved 2010-09-25.

[3] Greenall (1979) p.19

[4] Greenall (1979) p.20

[5] BBC - History - Tribes of Britain (http://www.bbc.co.uk/history/ancient/british_prehistory/iron_02.shtml). Retrieved 16 August 2009.

[6] Greenall (1979) p.29

[7] Wood, Michael (1986) *The Domesday Quest* p. 90, BBC Books, 1986 ISBN 0-563-52274-7.

[8] Mills, A.D. (1998). A Dictionary of English Place-names. Second Edition. Oxford University Press, Oxford. p256. ISBN 0-19-280074-4

[9] Rockingham Castle (http://www.rockinghamcastle.com/). Retrieved 16 August 2009.

[10] Mott, Allan. BBC - Cambridgeshire - History: Mary Queen of Scots' last days (http://www.bbc.co.uk/cambridgeshire/content/articles/ 2008/04/28/mary_queen_video_feature.shtml). Retrieved 16 August 2009.

[11] Stearns, Peter N., Langer. William L. The Encyclopedia of world history: ancient, medieval, and modern (http://books.google.co.uk/ books?id=MziRd4ddZz4C&pg=PA241&lpg=PA241&dq=Henry+VI+captured+Northampton&source=bl&ots=Y59GhdBeqO& sig=FVzhUUWK-UGWS-wwFgk1-BwMEI8&hl=en&ei=w3yISovTKJbUjAeS4qWiCw&sa=X&oi=book_result&ct=result& resnum=10#v=onepage&q=Henry VI captured Northampton&f=false). Retrieved 16 August 2009.

[12] The Writings of George Washington: Life of Washington (http://books.google.co.uk/books?id=PrfcVGtd0T4C&pg=PA545& lpg=PA545&dq=laurence+washington+warton&source=bl&ots=iyZ1YkQYH8&sig=I4ZUN9rFZ9eZwcNVgWY9rcwvNNY&hl=en& ei=cHqISprWEY3SjAf21YyiCw&sa=X&oi=book_result&ct=result&resnum=4#v=onepage&q=&f=false). Retrieved 16 August 2009.

[13] Edmonds. 1848. Notes on English history for the use of juvenile pupils (http://books.google.com/books?id=nCQEAAAAQAAJ& pg=PA49&dq=Charles+I+imprisoned+Holdenby&as_brr=3#v=onepage&q=&f=false). Retrieved 16 August 2009.

[14] Brookes, R., Whittaker, W.B. *The General Gazetteer, or, Compendious geographical dictionary, in miniature* (http://books.google.com/ books?id=XjANAAAAYAAJ&pg=RA2-PA241&dq=geography+of+northamptonshire&lr=&as_brr=1#v=onepage&q=&f=false). 1823. Retrieved 5 September 2009.

[15] Malte-Brun, C. Universal geography: or, A description of all parts of the world (http://books.google.com/books?id=-0gBAAAAYAAJ& pg=PA771&dq=geography+of+northamptonshire&as_brr=1#v=onepage&q=&f=false). 1832. Retrieved 5 September 2009.

[16] Andrews, R., Teller, M. The Rough Guide to Britain (http://books.google.co.uk/books?id=AOt1Hb8MOQUC&pg=PA553& lpg=PA553&dq=Northamptonshire+spires+and+squires&source=bl&ots=BjV7glaFoX&sig=JAOqwiJC80mHBAFb4qNFbQ8Jwqo& hl=en&ei=n56PSqO9DY-SjAePzoz4DQ&sa=X&oi=book_result&ct=result&resnum=8#v=onepage&q=&f=false) 2004. Rough Guides. Retrieved 5 September 2009.

[17] GENUKI: Northamptonshire Genealogy: Bartholomew's Gazetteer of the British Isles, 1887 (http://www.kellner.eclipse.co.uk/genuki/ NTH/). 11 August 2008. Retrieved 5 September 2009.

[18] Corby - English Partnerships (http://www.englishpartnerships.co.uk/corby.htm). Retrieved 16 August 2009.

[19] Northampton - English Partnerships (http://www.englishpartnerships.co.uk/northampton.htm). Retrieved 16 August 2009.

[20] Northamptonshire Chamber :: Milton Keynes & South Midlands Growth Plan (http://www.northants-chamber.co.uk/representation/ mksm/). Retrieved 16 August 2009.

[21] Peterborough Diocesan Registry (http://www.peterboroughdiocesanregistry.co.uk/). Retrieved 15 August 2009.

[22] The Huntingdon and Peterborough Order 1964 (SI 1964/367), see Local Government Commission for England (1958-1967), *Report and Proposals for the East Midlands General Review Area (Report No.3)*, 31 July 1961 and *Report and Proposals for the Lincolnshire and East Anglia General Review Area (Report No.9)*, 7 May 1965

[23] The English Non-metropolitan Districts (Definition) Order 1972 (SI 1972/2039) Part 5: County of Cambridgeshire

[24] Northamptonshire - Let yourself grow: Media information about Northamptonshire (http://www.explorenorthamptonshire.co.uk/exec/ 112345/7111/). Retrieved 15 August 2009.

[25] UK Genealogy Archives: Transcript from Pigot & Co's Commercial Directory, 1830 (http://uk-genealogy.org.uk/england/ Northamptonshire/pigot.html). Retrieved 15 August 2009.

[26] Northamptonshire Genealogy: Bartholomew's Gazetteer of the British Isles, 1887 (http://www.kellner.eclipse.co.uk/genuki/NTH/). Retrieved 15 August 2009.

[27] http://www.statistics.gov.uk/statbase/Expodata/Spreadsheets/D8271.csv

[28] Northamptonshire County Council: District and Borough Councils (http://www.northamptonshire.gov.uk/en/Pages/districts.aspx). 2008. Retrieved 22 August 2009.

[29] The Cambridgeshire (City of Peterborough) (Structural, Boundary and Electoral Changes) Order 1996 (http://www.legislation.gov.uk/ uksi/1996/1878/contents/made) (SI 1996/1878), see Local Government Commission for England (1992), *Final Recommendations for the Future Local Government of Cambridgeshire*, October 1994 and *Final Recommendations on the Future Local Government of Basildon & Thurrock, Blackburn & Blackpool, Broxtowe, Gedling & Rushcliffe, Dartford & Gravesham, Gillingham & Rochester upon Medway, Exeter, Gloucester, Halton & Warrington, Huntingdonshire & Peterborough, Northampton, Norwich, Spelthorne and the Wrekin*, December 1995

[30] Northamptonshire County Council: Members of Parliament (http://www.northamptonshire.gov.uk/en/councilservices/council/ mp_mep/pages/mps.aspx). 27 April 2009. Retrieved 22 August 2009.

[31] "Northamptonshire County Council website" (http://www.northamptonshire.gov.uk/). . Retrieved 4 June 2009.

[32] GENUKI: Northamptonshire Genealogy: Bartholomew's Gazetteer of the British Isles (http://www.kellner.eclipse.co.uk/genuki/NTH/). 1887. Retrieved 22 August 2009.

[33] Kellysearch.co.uk: R Griggs & Co. Ltd (http://www.kellysearch.co.uk/gb-company-370000597.html). Retrieved 22 August 2009.

[34] Northamptonshire Chamber: Major Northamptonshire employers (http://www.northants-chamber.co.uk/info/topemployers/#). Retrieved 22 August 2009.

[35] (http://www.wellingborough.gov.uk/downloads/Why_Wellingborough_Final_version.pdf). Retrieved 23 August 2009.

[36] Prologis RFI Dirft Daventry (http://www.prologisrfidirft.co.uk/). Retrieved 22 August 2009.

[37] FirstGBRf: FirstGBRf opens unique depot at Wellingborough (http://www.gbrailfreight.com/news.php?newsid=178). 12 June 2007. Retrieved 22 August 2009.

[38] GB Railfreight: Locations, Wellingborough (http://www.gbrailfreight.com/locations.php?lid=211) Retrieved 11 November 2010

[39] Regional Gross Value Added.*Office for National Statistics* (http://www.statistics.gov.uk/downloads/theme_economy/RegionalGVA. pdf). pp 240–253. 21 December 2005. Retrieved 22 August 2009.

[40] Components may not sum to totals due to rounding

[41] includes hunting and forestry

[42] includes energy and construction

[43] includes financial intermediation services indirectly measured

[44] Coe, N.M., Kelly, P.F, Wai-Chung Yeung, H. Economic geography: a contemporary introduction (http://books.google.co.uk/ books?id=1xV5ZvXYtjUC&pg=PA141&lpg=PA141&dq=motorsport+industry+northamptonshire&source=bl&ots=oM13gCrv3r& sig=RFRZCbtKwvSgU3QbDdGNtforJ6s&hl=en&ei=KlOQSqfCINyMjAfj8-HjDQ&sa=X&oi=book_result&ct=result& resnum=6#v=onepage&q=motorsport industry northamptonshire&f=false). Wiley-Blackwell, 2007. pp 141-143. Retrieved 22 August 2009.

[45] Russell Hotten. Motor racing battles to stay out of pits (http://business.timesonline.co.uk/tol/business/industry_sectors/leisure/ article5983082.ece). TimesOnline. 27 March 2009. Retrieved 22 August 2009.

[46] Official site of Mercedes GP Formula One Team: Contact us (http://www.mercedes-gp.com/includes/privacy.htm). Retrieved 4 March 2010.

[47] Force India F1 Team: Contact us (http://www.forceindiaf1.com/index/page_id/48). Retrieved 22 August 2009.

[48] Cosworth: Contact (http://www.cosworth.com/Default.aspx?id=1089541). Retrieved 22 August 2009.

[49] Mercedes-Benz High Performance Engines Ltd: Contact (http://www.mercedes-benz-hpe.com/hpe/index.htm). Retrieved 22 August 2009.

[50] Silverstone Official Website: Contact Numbers (http://www.silverstone.co.uk/php/ci_overview.html). Retrieved 22 August 2009.

[51] Getting to Rockingham (http://www.rockingham.co.uk/about/gettingto.asp). Retrieved 22 August 2009.

[52] Santa Pod Raceway: Contact/find us/postcode (http://www.santapod.co.uk/g_find.php). Retrieved 22 August 2009.

[53] Motorsport to grow 30% in next decade (http://www.northantset.co.uk/12691/Motorsport-to-grow-3037-in.5401811.jp). Northants Evening Telegraph. 25 June 2009. Retrieved 22 August 2009.

[54] MSKM: North Northants (http://www.mksm.org.uk/area/north-northants.asp) Accessed 2 October 2010

[55] MKSM: West Northants (http://www.mksm.org.uk/area/west-northants.asp) Accessed 2 October 2010

[56] Northants Evening Telegraph: Come to North Londonshire (http://www.northantset.co.uk/news/Come-to-North-Londonshire.6370328. jp) Accessed 2 October 2010

[57] North Londonshire: home page (http://www.northlondonshire.co.uk/) Accessed 2 October 2010

[58] Let yourself grow: home page (http://www.letyourselfgrow.com/) Accessed 2 October 2010

[59] Northamptonshire County Council: Northamptonshire Schools Directory (http://www3.northamptonshire.gov.uk/ncc/Templates/ content_applications.aspx?NRMODE=Published&NRORIGINALURL=/Learning/Institutions/schoolsdir. htm?SchoolDetail=9287031|Special%20-%20Primary&NRNODEGUID={7BF3FCE4-28B4-49B8-8F4E-CEA308D6E556}& NRCACHEHINT=NoModifyGuest). Retrieved 8 August 2009.

[60] Northamptonshire County Council: Saturday Music and Performing Arts Centres (http://www.northamptonshire.gov.uk/en/councilservices/EducationandLearning/music/Pages/sat_centres.aspx). Retrieved 8 August 2009.

[61] Northamptonshire County Council: Music Service: Youth Groups (http://www.northamptonshire.gov.uk/en/councilservices/EducationandLearning/music/Pages/YouthGroups.aspx). Retrieved 8 August 2009.

[62] Tresham College: Our Campuses (http://www.tresham.ac.uk/about/our_campuses). Retrieved 8 August 2009.

[63] Tresham College: Our Courses (http://www.tresham.ac.uk/courses). Retrieved 8 August 2009.

[64] Tresham College: Higher Education (http://www.tresham.ac.uk/higher_education). Retrieved 8 August 2009.

[65] The University of Northampton: About Us (http://www.northampton.ac.uk/about/). Retrieved 8 August 2009.

[66] The University of Northampton: Course finder (http://www.northampton.ac.uk/courses/search/). Retrieved 8 August 2009.

[67] "New £4.2m Irthlingborough outpatients clinic opens" (http://www.bbc.co.uk/news/uk-england-northamptonshire-12379676). BBC News. 7 February 2011. . Retrieved 7th February 2011.

[68] Tite, Nick (2008-07-14). "Rabbit caused water contamination at Pitsford - Northants ET" (http://www.northantset.co.uk/news/Rabbit-caused-water-contamination-at.4286344.jp). Northants Evening Telegraph. . Retrieved 2008-08-22.

[69] "Sickness bug found in tap water" (http://news.bbc.co.uk/1/hi/england/northamptonshire/7472619.stm). BBC. 2008-06-25. . Retrieved 2008-07-15.

[70] "BBC News". *News at Ten, BBC One* (BBC). 2008-07-14.

[71] "Anglian Water" (http://www.anglianwater.co.uk/index.php?sectionid=51&parentid=50&contentid=1136), Press Release

[72] "Water crisis: All clear for tap water - and up to £30 compensation! - Northampton Chronicle and Echo" (http://www.northamptonchron.co.uk/news/25-each-compensation-for-water.4255069.jp). Chronicle & Echo. 2008-07-05. . Retrieved 2008-08-22.

[73] Britten, Nick (2009-02-23). "Corby station" (http://www.telegraph.co.uk/news/uknews/road-and-rail-transport/4787477/The-17million-train-station-with-only-one-service-a-day.html). London: Telegraph.co.uk. . Retrieved 2010-09-25.

[74] Network South East routes (http://www.nsers.org.uk/nse2.htm)

[75] "SMJR" (http://www.smjr.info). Smjr.info. 2010-09-19. . Retrieved 2010-09-25.

[76] (http://www.atoc.org/general/ConnectingCommunitiesReport_S10.pdf) ATOC Connecting Communities Report

[77] Network Rail: East Midlands Draft Route Utilisation Strategy (http://www.networkrail.co.uk/browse documents/rus documents/route utilisation strategies/east midlands/east midlands rus draft for consultation.pdf) Access date: 4th January 2010]

[78] Stagecoach Northants (http://www.stagecoachbus.com/northants/index.html)

[79] "First Northampton: Timetables" (http://www.firstgroup.com/ukbus/eastmidlands/northampton/timetables/index.php?operator=20&page=1&redirect=no). Firstgroup.com. 2010-09-19. . Retrieved 2010-09-25.

[80] http://www.ead.eurocontrol.int/eadbasic/pamslight-113B93979B7148E51E98BD22C3FF290E/7FE5QZZF3FXUS/EN/AIP/AD/EG_AD_2_EGBK_en_2011-03-10.pdf

[81] http://www.theroadmender.com/

References

• Greenall, R. L. (1979) *A History of Northamptonshire* Phillimore & Co. Ltd. ISBN 1-86077-147-5.

External links

• Northamptonshire County Council (http://www.northamptonshire.gov.uk/en/Pages/HomePage.aspx)

• Northamptonshire (http://www.dmoz.org/Regional/Europe/United_Kingdom/England/Northamptonshire/) at the Open Directory Project

• Northamptonshire Images and Information (http://www.northamptonshire.co.uk/)

• 1894/5 description (http://www.uk-genealogy.org.uk/gazetteer/england/Northamptonshire/)

• Northants Forum (http://northantscommunitycafe.co.uk/)

• Local Theatre in Northamptonshire (http://www.theatrenights.com/index.php?list=Events&location=4)

• Northamptonshire History Website (http://www.northamptonshire-history.org.uk/)

• Northamptonshire Tourism Website (http://www.explorenorthamptonshire.co.uk/)

• Northamptonshire Guide Website (http://www.northamptonshireguide.co.uk/)

• Visit Northamptonshire Website (http://www.visitnorthamptonshire.co.uk/)

• Northamptonshire Online Forum (http://www.northamptonshireforum.co.uk/)

• NorthantsSavings.com (Northamptonshire deals) (http://www.northantssavings.com/)

Article Sources and Contributors

Brington,_Northamptonshire *Source*: http://en.wikipedia.org/w/index.php?title=Brington%2C_Northamptonshire *Contributors*: Bobschops, Bogbumper, Daemonic Kangaroo, Grutness, Lozleader, Saga City, Steinsky, Uksignpix, 1 anonymous edits

Wanderers_F.C. *Source*: http://en.wikipedia.org/w/index.php?title=Wanderers_F.C. *Contributors*: A18919, ADL1983, Airwolf, Apanuggpak, Arbero, Arwel Parry, Bartash, Bearcat, Ben davison, Bissinger, BlackJack, Bornintheguz, Casliber, Chaz1dave, ChrisTheDude, Clio64B, Dabomb87, Daemonic Kangaroo, Dispenser, Djln, DrFrench, Eddie6705, Erik Kennedy, Etacar11, Everton, Felipe P, Footballwecan80, Foxhill, GO, GiantSnowman, Graham87, Greenshed, Grunners, Hartmann Linge, IceDragon64, Jenks24, Jnestorius, JustAGal, KJPurscell, Kauczuk, Kinigi, Kisbie, Kobrabones, Kvaratschelia, Kwib, Leszek Jańczuk, Malpass93, Mikedash, Nzd, Od Mishehu, Osomec, PakistanGangster, Paxton Pete, Penlid, Qwghlm, Rambo's Revenge, Reinhardheydt, Rugby monk, Sarumio, Siva1979, Sjorford, Slumgum, Somersetccc, Spitfire, Starfighter Pilot, Surge79uwf, The wub, Thumperward, Tim!, Ukmarkwilson, WhatGuy, ‏יולוקורב‎, 40 anonymous edits

Daventry_District *Source*: http://en.wikipedia.org/w/index.php?title=Daventry_District *Contributors*: Acalamari, Aranel, AxG, Barryob, Bobo192, Bobschops, Cnyborg, Dallan72, DaventryTownClerk, Francs2000, G-Man, Gardar Rurak, Gralo, Guthrie, Haydnaston, JBellis, Keith Edkins, Likelife, Lozleader, MRSC, Maximus Rex, Morwen, Nilfanion, Oliver Chettle, Peter Shearan, Plastikspork, ReformatMe, Rich Farmbrough, Saga City, Senator Palpatine, SilkTork, SteveMtl, Template namespace initialisation script, Uksignpix, Winston365, 6 anonymous edits

Henry_Holmes_Stewart *Source*: http://en.wikipedia.org/w/index.php?title=Henry_Holmes_Stewart *Contributors*: Daemonic Kangaroo, MBisanz, Timrollpickering, WOSlinker, Waacstats

Nobottle *Source*: http://en.wikipedia.org/w/index.php?title=Nobottle *Contributors*: Auric, Bearcat, Bogbumper, Brookie, Dave.Dunford, Icerat, LilHelpa, PJM, Saga City, SchuminWeb, Stavros1, Storye book, Verica Atrebatum, 3 anonymous edits

Rector *Source*: http://en.wikipedia.org/w/index.php?title=Rector *Contributors*: 8472, Abach, Acategory, Adambisset, Adrian two, Afasmit, Agemegos, Alai, Alan16, Alexf, Amortize, AndreSt, Andreas Kaganov, Andres, Andycjp, Angusmclellan, Anthropax, Armindo, Arnoutf, Awbeal, BD2412, Bcwright, Bgohla, Bhuck, Big Jim Fae Scotland, Bjankuloski06en, Bluerfn, Bob Burkhardt, Bomac, BozoTheScary, Breadandcheese, Bryanmaguire, Calabe1992, CanisRufus, Carlesmari, Carolynparrishfan, Cd94, Cherkash, Chester Markel, Chill doubt, Chipmunker, ChristopheS, Cimmerian praetor, Cocytus, Colin S, Concledoc, Corvokarasu, Ctill, DGG, DJ Bungi, Dad Buck, Dave souza, David Shay, Davidkinnen, Dbracken3, Doops, DorAlbi, Dracunculus, Drawdennep, Dumfie, Ectoras, Eddaido, Eddieuny, Emaha, Epanalepsis, Epbr123, Evans1982, Fastifex, Fayenatic london, Fishhead64, Fishiehelper2, Francis Davey, Future Perfect at Sunrise, Gaius Cornelius, Gary D, Gentgeen, Gerry Lynch, Glenn, Gokhansayram, Grievous Angel, Gumruch, H2ppyme, HHahn, HarvardOxon, Henry Flower, Hhbruun, Hmains, Htonl, Ihcoyc, Ilya-42, Inwind, Iridescent, Irishguy, Ivan Chernyenko, Ixnayonthetimmay, JASpencer, Jarmo Gombos, Jastcy, Jfdwolff, Jiang, Jlittlet, Jnetusil, John, Joy, Jpbowen, Jsernest, Jtdirl, Karen Johnson, Ken Gallager, Kevin Rector, Khazar2, Kilbosh, KnightRider, Licor, LilHelpa, Lincolnite, Lisatwo, Littlealien182, Lordrosemount, Lstanley1979, Lucifero4, Lupo, MB-one, Macdonald-ross, Mais oui!, Malyctenar, Mangoe, Marcetw, Marcok, Michael Hardy, Michael riber jorgensen, Micru, Mlaffs, Momet, Nein7, Nejee16, Nihil novi, Nikola Smolenski, Nv8200p, OJH, OlEnglish, OneNomad, Oosoom, Op. Deo, Orangemike, Orenburg1, Oxymoron01, PBS-AWB, Page Up, PainMan, Palthrow, Parkwells, PatGallacher, Paterm, Pavel Vozenilek, Pcgomes, Pdfpdf, Peter Bell, Peterlin, PetiteFadette, Pgg7, Philippe Giabbanelli, Philtro, Physicistjedi, Piledhigheranddeeper, Pilot320, Pj44300, Pmadrid, Pudeo, R'n'B, Ragesoss, RaySys, Rcollins8, Reinis, Rgvis, Richard Keatinge, Rimakela, Rmbyoung, Rockhopper10r, Rotring, Rrius, Santryl, SchuminWeb, Sethemanuel, Skier Dude, Skittleys, Sky89, Skysmith, Slackermonkey, Sluzzelin, Spaceman85, Station1, Sterio, Tenmei, Therepel, Timrollpickering, UpDown, Urednik, Urli mancati, Vanished user 03, Vaquero100, Vegfarandi, Venerock, Vervin, Vgy7ujm, Victoriaedwards, Vikslen, Vizcarra, Volunteer Marek, Westleigh T.Br, Wigert, Winterst, Wolfman3219, Zacheus, Александр Мотин, 247 anonymous edits

Great_Brington *Source*: http://en.wikipedia.org/w/index.php?title=Great_Brington *Contributors*: 1oddbins1, AmosWolfe, AvicAWB, Bogbumper, Gatechjon, Grutness, Lozleader, NinetyCharacters, Proteus, Reedy, Saga City, Sir Stanley, Steinsky, 12 anonymous edits

FA_Cup *Source*: http://en.wikipedia.org/w/index.php?title=FA_Cup *Contributors*: 'Arry Boy, *Paul*, 03md, 2007phon, 4000holes, A.K.A.47, AB-me, ABF, AFrayn, Aaron carass, Aaronhumes, Aboutmovies, Achangeisasgoodasa, Acsenray, Adamjhepton, Adzz, Aecis, AfC, Afterschoolworld, Ahoerstemeier, Ajuk, Akrabbim, Alai, Alansohn, Aleaf.biz, AlexWilkes, Alias Flood, Alii h, Allstarecho, Alongseptember, Alphax, Amcw7760, Andi kan, Andy Smith, AndyMooney, Andyboi999, Angmering, Anthony R, Apstockholm, Arniep, Arsenal Are the dogs nuts, Art LaPella, Arthena, Arwel Parry, Ary29, Athosfolk, BDBJ, BGTopDon, Bababoum, Bad attitude, Badmronaki, BaldBoris, Bamf 999, BanRay, Barald316, Barryap, Bbogdanleo, Bdsr, Beach bum7, Bebofpenge, Beckford14, Bedders, Ben davison, Benjamus, BernardBrowne, Beyond495, Bigmike, Bilsonius, Binabik80, Birchy007, Bjf, Black Falcon, BlackbeltMage, Blacklung08, Blonde Ashcroft, Bluelion, Bob Palin, Bobbymozza, Bobo192, Boing! said Zebedee, Bornintheguz, Brad78, Brandmeister (old), Branson03, Brickie, Burgs501, Bwfcdan, CO, Caitlin Hanks, CalJW, Callipides, Calsicol, CambridgeBayWeather, Camembert, Camerong, Can't sleep, clown will eat me, CanadianLinuxUser, Cardiocortez, Catchpole, Celticosprey, Centrx, Chandler, Chanheigeorge, Chappy84, Chasetown07, Cheeseklaxons, Chivista, Chosen John, Chris the speller, ChrisTheDude, Chrislintott, Chrism, Chrissii-tina, ChristalPalace, Christopher Connor, Chubby Gazelle, Chuunen Baka, Ckatz, Cleverjanos, Cmlau, Coelacan, CommonsDelinker, Conversion script, Craptacular, Crookesmoor, Cs-wolves, Cst17, Cutler, Cyranodeboston, Dabomb87, Daemonic Kangaroo, Dale Arnett, DandyDan2007, Danielk2, Danmantalis, Danread, Darryl.matheson, Dave w74, DaveB21, DaveJB, Daveb, Davelewis, David Thrale, Dawn Bard, DePiep, Deanceltics, Dekisugi, Delusion23, Diddylevine, Differentgravy, Difgbdfgioudbsfgdlfiugbfgpdfng, Dirk Valentine, Discospinster, Djdaedalus, Djln, Dkua, Dmhobbs, Dmontin, Down-in-our-albion, Drc79, Drumncars1996, Dudesleeper, Duncharris, Dunnville, EJF, EPs, EamonnPKeane, Ebessan, Ed g2s, Edderso, Edgar181, Edmund Blackadder, Edward, Egghead06, Elliskev, Emtee65, Epistemophiliac, Ericamick, Erik9, Erzengel, Euroleague, Evercat, Fatla00, Feinoha, Feudonym, Fiddle and herman, Finns, Fireburn95, Fish1987, Fish1989, Fjmustak, Flix11, Flying porkpie, Footballexpert, Forbsey, Foxj, Frecuentaje, Freddieandthedreamers, Fredil Yupigo, Fryede, Gabbec, Gadfium, Gaius Cornelius, Gareth E Kegg, Garion96, Gasheadsteve, Gazzastag, Geboy, Georgeg, Ging Ning, GoldDragon, Golgofrinchian, Goodie00, Gooner1990, Goten39, Graham87, GrahamHardy, Grandsorbs, Green Tentacle, Gregorybean, Grunkhead, Grutness, Gwdr500, Gypsum Fantastic, Hairy Dude, Halmstad, Hammersfan, Handballdemigod, Haroldsomers, Harrias, HarryCoombes, Harvey'swiki, Hawksworthm, Hede2000, Henrygb, Herandar, Heryu, Hibou8, Hillel, HonorTheKing, Howardmcn, Hubsafcw, Hux, Hvn0413, HyuenK, ISD, ITVL1972, Icairns, Illuminattile, Illuminaut, Ingumsky, J.P.Lon, JB82, JFBridge, JHK, JPH-FM, JRRobinson, JT72, Jae.oh, Jameboy, JamesWikip, Jarv1178, Jcarls1, Jclemens, JeremyA, Jeronimo, Jfdjfd, JiVE, Jim1138, Jimbo online, Jklin, Jnestorius, Jockosaurus, JoeC 4321, JoeWiki, Joebamber1990, JoelCFC25, John, John wesley, Johnatx, Jojomcdowell, Jooler, JordanRD, Jordzmcguire, Joshurtree, Josquius, Juggling john, Kabads, Kajervi, Kanabekobaton, Kanaye, Keilana, Keith D, Ketiltrout, Kevin McE, Kiewzhenyi, KingStrato, Kingfisherswift, Kingk21, Kingturtle, Kittybrewster, Koavf, Krob74, Kronix1986, Kroome111, KyleRGiggs, L Kensington, Lachlanrulez, Lafuzion, Laslovarga, Latics, Lawrennd, Lburdon, Leaky caldron, Leberquesgue, Leffrin, Leinvisiblegarcon, LemonGrass68, Leszek Jańczuk, Lexicon, Lfclad007, Lightmouse, Lights, LilHelpa, Lion King, Loganberry, Lord Jubjub, Lukasa, LukeSurl, Lukedavidwallis, Lutton, MC MasterChef, MLD, MOTORAL1987, MSJapan, MTC, Maarten1963, Mabuska, Macaulay24, Macy, MadSproute, Madw, Mais oui!, Malcolmxl5, Malpass93, Manchester10, Mandel, Manofedit, Manop, MapsMan, Mark272, MartinUK, Matfo, Mathsguy2003, Matt91486, Matthew hk, MatthewMain, Matthewmayer, Matthijs2006, Maxim, Mbr25, McChelski, Mcauburn, Mcoady12, Megastar, Mintguy, Mirv, MisfitToys, Mohsenkazempur, Morwen, Mpbx3003, Mr Hall of England, MrH, Muppet, Musicandnintendo, Musungu jim, Mymap, N5iln, Nairobiny, Navidff, Neier, Nevilley, Nez202, Niall123, Nick C, Nickbiebuyck, Nitsansh, North wales cestrian, Novadeath69, NuclearWarfare, Number 57, Nutmeg25, Nygoodliving, Oalexander-En, Oberonfoxie, Odie5533, Oldelpaso, Ommlette, One Salient Oversight, Onebravemonkey, Optimizerone, Orphan Wiki, Owainjohnevans, Paalegge, Pakpoom da, Pal, Palmiped, PalmyPete, Pasadena, Patrick Neylan, Paul-L, Paulbrock, Paulwmk, Peanut4, PeeJay2K3, Peidu, Pelmeen10, Peoplesunionpro, Phantlers, Phantomsnake, Pharaoh of the Wizards, Phildav76, Philip Trueman, Phillper0906, Php2rh, Physiology, Piano non troppo, Picapica, Pjibid, Pobbie Rarr, Pretty Green, Pricey3000, Prin S, Quadalpha, Quentin X, Quicksilvre, Qwghlm, R Lowry, R'son-W, RE LFC93, RFBailey, RaptorX, Rayshawndiggs, Rbrwr, Rdalvie, Reconsider the static, RednessInside, Refsworldlee, Regan123, Retinarow, RexNL, Rich Farmbrough, Richard Rundle, Richard W.M. Jones, Rik201, Rising*From*Ashes, Ritchie41, Rjwilmsi, Rmnmn, Rngould91, Robdurbar, Robertgreer, RobinCarmody, Robwingfield, RohypnolDream, RomeW, Ronhjones, Ronvelig, Roserex57, Ross1, Rpisces11, Rsly93, Rup235, Rupertslander, Salmon89, Salttheearth, Sam Pointon, Samrobbo2, Sandybag, SantosOctagon, Saracen1970, Sars, Scorchy, Seadog365, Sean4Lancer, Secretlondon, Seedybob2, Seemo82, SenorKristobbal, Sephjnr, Seraphim, Setantaaustralia, Seth Whales, Severo, Sf07, Shambodalton, SiOfUmbar, SilkTork, Sillyfolkboy, Similaun, SimonMayer, Siva1979, Sjorford, Slumgum, Snowmanradio, SpLoT, Spdwolf, Speedking90, Spiderlounge, Sport9, Srikeit, Starbois, Starfighter Pilot, Startstop123, Statto74, Statto999, Stephen MUFC, Stephenb, Steve Farrell, SteveO, Stevebritgimp, Stevew2022, Steveweiser, Stevo1000, Stifle, Struway, Struway2, Suicidalhamster, Sunil060902, Surge79uwf, TBM10, THELEICESTERFOX, TN2006, Talkshowbob, TattooedLibrarian, Tcncv, TekhneGrammatike, Terence, Testing times, Thburrow, The C of E, The Font, The Frederick, The Rambling Man, The Tramp, The wub, TheJC, Theknightswhosay, ThirdEdition, Thumperward, Tide rolls, Tim!, Tlusfa, Tmopkisn, Tocino, Tolkny, Tommytumer7, Tompw, Tony2Times, Tonywalton, ToonIsALoon, Topperfalkon, Trieste, Trombles, Tryde, Trödel, Twas Now, Ukdazza, Unisouth, Unreal7, V4R, VEO15, Valenciano, Venatoreng, Villafanuk, Vyom1996, WATP, Warofdreams, Warut, Watty1962, Wedgehedge444, Wehwalt, Wembwandt, Whitecyda, Wikiadms, Wikien2009, Wikipelli, William Avery, WilliamF1two, Willy on WheeLs, Wjemather, Wjemother, Wrighty1903, X96lee15, Xenomorph1984, Xlobyx1, Yorkshiresky, Yuui5, Z2a, ZAccelerator, Zaphod Beeblebrox, Zipgun, Zuhdizuhdi, Zzyzx11, Zé da Silva, Рядинский Евгений, , 1376 anonymous edits

Northamptonshire *Source*: http://en.wikipedia.org/w/index.php?title=Northamptonshire *Contributors*: 1jord2, 80.255, Aaronjhill, Acalamari, Ahoerstemeier, Akadruid, Al Silonov, Alansohn, Amazonien, Ameliorate!, AmosWolfe, Andres, Andrew Dalby, Andrew Norman, Andycjp, Angr, AnnaFrance, AnnaP, Anwar saadat, AssociateAffiliate, Autoerrant, Bcasterline, Bigger digger, Birchington, Bleaney, Bogbumper, Bornintheguz, Brandon, Brookie, Brool, BrownHairedGirl, Bryan Derksen, Bwithh, CLW, Caponer, Capricorn42, Carlwev, CarolGray, Chanheigeorge, Charles Matthews, Childzy, Chris the speller, Chrisieboy, Citterio, Cj1340, Clear air turbulence, Clio64B, Cmdrjameson, Cnyborg, Cphaff, Craigsteele2001, D6, DH85868993, Dale Arnett, Dallan72, DaventryCobbler, Deflective, Delusion23, Deror avi, Deville, DinosaursLoveExistence, Discospinster, Dn9ahx, DoctorW, Docu, Dpaajones, DrKiernan, Duncan Keith, Dusimpson, Ehrenkater, Eopsid, Epbr123, Erik Kennedy, Flatterworld, Foreverprovence, Fourohfour, Francs2000, Fuhghettaboutit, G-Man, Gaius Cornelius, Gene Nygaard, Giano, Grev02, Ground Zero, Grstain, Gsingh12345, Hankwang, Happy Radio, HennessyC, Heron, Hiram K Hackenbacker, Howard Alexander, Ilikeeatingwaffles, Iloveaqua, Interplanet Janet, Iridescent, Istw, Iulianu, JBellis, Jamesinderbyshire, Jamie Mercer, Jatrius, Jeffthejiff, Jeni, Jim 21, Jimbo online, Jimly9, Jimscotland, Jjustyy, Jklin, John Maynard Friedman, John of Reading, Joshua Issac, KathrynLybarger, Keith D, Keith Edkins, Ketiltrout, Koavf, Kubigula, Kudpung, Kummi, Kwamikagami, Lampman, Lang rabbie, Laurel Lodged, Le Creillois, Likelife, LilHelpa, Lisagosselin,

Image Sources, Licenses and Contributors

file:Northamptonshire UK location map.svg Source: http://en.wikipedia.org/w/index.php?title=File:Northamptonshire_UK_location_map.svg License: unknown Contributors: User:Nilfanion

File:Red pog.svg Source: http://en.wikipedia.org/w/index.php?title=File:Red_pog.svg License: unknown Contributors: Anomie

File:Wanderers.gif Source: http://en.wikipedia.org/w/index.php?title=File:Wanderers.gif License: unknown Contributors: User:Ukmarkwilson

Image:ForestFC1863.jpg Source: http://en.wikipedia.org/w/index.php?title=File:ForestFC1863.jpg License: unknown Contributors: Original uploader was ChrisTheDude at en.wikipedia

Image:QPWanderers.jpg Source: http://en.wikipedia.org/w/index.php?title=File:QPWanderers.jpg License: unknown Contributors: Cynec

File:Modern Wanderers Team.jpg Source: http://en.wikipedia.org/w/index.php?title=File:Modern_Wanderers_Team.jpg License: unknown Contributors: User:Ukmarkwilson

Image:1896 FA Cup.jpg Source: http://en.wikipedia.org/w/index.php?title=File:1896_FA_Cup.jpg License: unknown Contributors: User:Oldelpaso

File:Daventry UK locator map.svg Source: http://en.wikipedia.org/w/index.php?title=File:Daventry_UK_locator_map.svg License: unknown Contributors: User:Nilfanion

Image:Townsend Farm, Nobottle - geograph.org.uk - 130518.jpg Source: http://en.wikipedia.org/w/index.php?title=File:Townsend_Farm,_Nobottle_-_geograph.org.uk_-_130518.jpg License: unknown Contributors:

File:Inauguration-Lubomir-Dvorak.jpg Source: http://en.wikipedia.org/w/index.php?title=File:Inauguration-Lubomir-Dvorak.jpg License: unknown Contributors: User:snek01, User:snek01

Image:YushimaSeido8678.jpg Source: http://en.wikipedia.org/w/index.php?title=File:YushimaSeido8678.jpg License: unknown Contributors: Dogears, Fg2, J o, Père Igor, Reggaeman

File:wikisource-logo.svg Source: http://en.wikipedia.org/w/index.php?title=File:Wikisource-logo.svg License: unknown Contributors: Nicholas Moreau

File:Thefacup-logo.png Source: http://en.wikipedia.org/w/index.php?title=File:Thefacup-logo.png License: unknown Contributors: Il223334234, Malpass93, Scorchy, TheBigJagielka

File:Flag of England.svg Source: http://en.wikipedia.org/w/index.php?title=File:Flag_of_England.svg License: unknown Contributors: Anomie

Image:Soccerball current event.svg Source: http://en.wikipedia.org/w/index.php?title=File:Soccerball_current_event.svg License: unknown Contributors: User:Anomie, User:Davidgothberg, User:Pumbaa80

File:1896 FA Cup.jpg Source: http://en.wikipedia.org/w/index.php?title=File:1896_FA_Cup.jpg License: unknown Contributors: User:Oldelpaso

File:The FA Cup Trophy in 2008.jpg Source: http://en.wikipedia.org/w/index.php?title=File:The_FA_Cup_Trophy_in_2008.jpg License: unknown Contributors: User:Unisouth

File:County Flag of Northamptonshire.png Source: http://en.wikipedia.org/w/index.php?title=File:County_Flag_of_Northamptonshire.png License: unknown Contributors: User:Jza84

File:Northamptonshire UK locator map 2010.svg Source: http://en.wikipedia.org/w/index.php?title=File:Northamptonshire_UK_locator_map_2010.svg License: unknown Contributors: User:Nilfanion

File:Coat of arms of Northamptonshire County Council.png Source: http://en.wikipedia.org/w/index.php?title=File:Coat_of_arms_of_Northamptonshire_County_Council.png License: unknown Contributors: User:Gaeser, User:Jza84, User:Poznaniak, User:Sodacan, User:Xavigivax

Image:NorthamptonshireNumbered.png Source: http://en.wikipedia.org/w/index.php?title=File:NorthamptonshireNumbered.png License: unknown Contributors: Michiel1972, Nilfanion, Skinsmoke

File:Loudspeaker.svg Source: http://en.wikipedia.org/w/index.php?title=File:Loudspeaker.svg License: unknown Contributors: Bayo, Gmaxwell, Husky, Iamunknown, Mirithing, Myself488, Nethac DIU, Omegatron, Rocket000, The Evil IP address, Wouterhagens, 18 anonymous edits

Image:Speed Northampton.jpg Source: http://en.wikipedia.org/w/index.php?title=File:Speed_Northampton.jpg License: unknown Contributors: Original uploader was Brookie at en.wikipedia

File:Orange pog.svg Source: http://en.wikipedia.org/w/index.php?title=File:Orange_pog.svg License: unknown Contributors: Andux, Antonsusi, Juiced lemon, Roomba, Sevela.p, TwoWings, Wgabrie, Wst

Image:Kilworth Wharf - geograph.org.uk - 164606.jpg Source: http://en.wikipedia.org/w/index.php?title=File:Kilworth_Wharf_-_geograph.org.uk_-_164606.jpg License: unknown Contributors: Auntof6

File:Kimi Raikkonen 2006 test.jpg Source: http://en.wikipedia.org/w/index.php?title=File:Kimi_Raikkonen_2006_test.jpg License: unknown Contributors: BrianScott

Image:A43 Brackley.jpg Source: http://en.wikipedia.org/w/index.php?title=File:A43_Brackley.jpg License: unknown Contributors: Adambro, Scillystuff, Seth Whales

Image:Grand Union Canal at Braunston.jpg Source: http://en.wikipedia.org/w/index.php?title=File:Grand_Union_Canal_at_Braunston.jpg License: unknown Contributors: Original uploader was G-Man at en.wikipedia

Image:Next stop Wellingborough - geograph.org.uk - 1400370.jpg Source: http://en.wikipedia.org/w/index.php?title=File:Next_stop_Wellingborough_-_geograph.org.uk_-_1400370.jpg License: unknown Contributors: John Webber

Image:Sywell Aerodrome.jpg Source: http://en.wikipedia.org/w/index.php?title=File:Sywell_Aerodrome.jpg License: unknown Contributors: delusional

Image:Broadcasting House, Northampton.jpg Source: http://en.wikipedia.org/w/index.php?title=File:Broadcasting_House,_Northampton.jpg License: unknown Contributors: . Original uploader was Tom- at en.wikipedia

Image:Bronze Statue Northampton RFC.jpg Source: http://en.wikipedia.org/w/index.php?title=File:Bronze_Statue_Northampton_RFC.jpg License: unknown Contributors: Michael Trolove

Image:AP_Icon.svg Source: http://en.wikipedia.org/w/index.php?title=File:AP_Icon.svg License: unknown Contributors: User:Victovoi

Image:UKAL icon.svg Source: http://en.wikipedia.org/w/index.php?title=File:UKAL_icon.svg License: unknown Contributors: User:Beao, User:Marknew

Image:Themepark uk icon.png Source: http://en.wikipedia.org/w/index.php?title=File:Themepark_uk_icon.png License: unknown Contributors: User:Hogweard

Image:CL_icon.svg Source: http://en.wikipedia.org/w/index.php?title=File:CL_icon.svg License: unknown Contributors: WebHamster. Original uploader was at

File:Country parks.svg Source: http://en.wikipedia.org/w/index.php?title=File:Country_parks.svg License: unknown Contributors: Kameraad Pjotr, S19991002, Sfan00 IMG, Svgalbertian, Tafkam

Image:EH icon.svg Source: http://en.wikipedia.org/w/index.php?title=File:EH_icon.svg License: unknown Contributors: WebHamster. Original uploader was at

Image:Forestry commission logo.svg Source: http://en.wikipedia.org/w/index.php?title=File:Forestry_commission_logo.svg License: unknown Contributors: AnRo0002, S19991002, Segu

Image:HR icon.svg Source: http://en.wikipedia.org/w/index.php?title=File:HR_icon.svg License: unknown Contributors: Bayo, Beao

Image:HH icon.svg Source: http://en.wikipedia.org/w/index.php?title=File:HH_icon.svg License: unknown Contributors: User:Beao

Image:Museum icon.svg Source: http://en.wikipedia.org/w/index.php?title=File:Museum_icon.svg License: unknown Contributors: User:Beao

Image:Museum icon (red).svg Source: http://en.wikipedia.org/w/index.php?title=File:Museum_icon_(red).svg License: unknown Contributors: User:Beao

Image:NTE icon.svg Source: http://en.wikipedia.org/w/index.php?title=File:NTE_icon.svg License: unknown Contributors: WebHamster. Original uploader was WebHamster at en.wikipedia

File:Drama-icon.svg Source: http://en.wikipedia.org/w/index.php?title=File:Drama-icon.svg License: unknown Contributors: User:Booyabazooka

File:Zoo icon.jpg Source: http://en.wikipedia.org/w/index.php?title=File:Zoo_icon.jpg License: unknown Contributors: Original uploader was Williams119 at en.wikipedia

Image:Museum icon (red).png Source: http://en.wikipedia.org/w/index.php?title=File:Museum_icon_(red).png License: unknown Contributors: Beao, Burts, CyberSkull, Duesentrieb, EugeneZelenko, Koavf, Marknew, Paddy

Image:HH icon.png Source: http://en.wikipedia.org/w/index.php?title=File:HH_icon.png License: unknown Contributors: Beao, CyberSkull, Duesentrieb, EugeneZelenko, Koavf, Marknew, Paddy, 2 anonymous edits

Image:CP icon.png Source: http://en.wikipedia.org/w/index.php?title=File:CP_icon.png License: unknown Contributors: Beao, Duesentrieb, EugeneZelenko, Koavf, Marknew, Paddy, Perhelion

Image:NTE icon.png Source: http://en.wikipedia.org/w/index.php?title=File:NTE_icon.png License: unknown Contributors: FSII, Grenavitar, Grstain, Sfan00 IMG, Steinsky, The Pink Oboe, WebHamster

Image:UKAL icon.png Source: http://en.wikipedia.org/w/index.php?title=File:UKAL_icon.png License: unknown Contributors: Beao, CyberSkull, Duesentrieb, EugeneZelenko, Koavf, Marknew, Paddy

Image:FC icon.png Source: http://en.wikipedia.org/w/index.php?title=File:FC_icon.png License: unknown Contributors: Original uploader was Steinsky at en.wikipedia

Image:Museum icon.png Source: http://en.wikipedia.org/w/index.php?title=File:Museum_icon.png License: unknown Contributors: Beao, Burts, CyberSkull, Duesentrieb, EugeneZelenko, Koavf, Marknew, Paddy

Printed by Books on Demand GmbH, Norderstedt / Germany

Brington, Northamptonshire

Please note that the content of this book primarily consists of articles available from Wikipedia or other free sources online. Brington is a civil parish in the Daventry district of the county of Northamptonshire in England. At the time of the 2001 census, the parish population was 482 people.It contains three villages:Great Brington, Little Brington, Nobottle. Office of National Statistics: Brington CP: Parish headcounts. Retrieved 7 November 2009. Venn, J.; Venn, J. A., eds. (1922–1958). "Stewart, Henry". Alumni Cantabrigienses (10 vols) (online ed.). Cambridge University Press.

978-613-6-25997-0